tp1.3

A continuing inquiry
into the
Foundations of the Science of Physics:

Vector Calculus I

JRBreton

tp1.3
A continuing inquiry into the
Foundations of the Science of Physics:
Vector Calculus I

by JRBreton

Published by:
The Foundation for Theoretical Physics 3 Apple Tree Lane
Walpole, MA 02081−2301

Web address: FoundationForTheoreticalPhysics.org
email address: theoretical.physics.books@gmail.com

Copies of this book and other offerings of the Foundation may be obtained preferably online from tp.vendevor.com, the Foundation's website store and also from amazon.com and similar sites.

ISBN, print ed. 978-0-9844299-9-8

First printing 2021

Printed in the United States of America

Library of Congress Control Number: 2021921167

Table of Contents

The Next Day

The morning of the next day found the three friends, Newton, Einstein, and Breton, well breakfasted, in their clubhouse, seated by the fire in their comfortable Windsors, each harboring a tempered eagerness to continue their investigation of Theoretical Physics. The smell of Autumn danced again with the sound of a log crackling in the fireplace. The room was no less cozy and peaceful; warm but not hot; lighted, but not bright.

Newton, who loved to summarize and recapitulate as well as construct tables, began the conversation with an attempt to summarize the conclusions of the previous days. "We agreed that a science like Physics differs from a technology like Surveying and that sciences differ from each other too. The science of Physics differs from the science of Mathematics. We accepted, after some debate the following definition of Physics:

Physics is the study of reality
observable as extended, moving, or forcing.

We concluded that the symbols and ideas of Mathematics were inappropriate for the science of Physics because they easily lead to ambiguities and false conclusions. We saw how the abuse of language and symbols becomes treacherous for anyone seeking the truth in any science. Since truth in general and scientific truth in particular cannot tolerate contradictions, sciences are forced to construct special dictionaries to avoid ambiguities in their disciplines. Mathematics has its dictionary; Physics should also have its own, distinct from the one for Mathematics. We call the dictionary for the science of Physics by the name **Theoretical Physics**. It, not Mathematics, is the proper language for Physics."

Breton: "So we embarked on a great adventure to discover how mathematical ideas and propositions can be transformed into ideas and propositions suitable for Theoretical Physics. The outlines of the adventure are clear enough: to transform any mathematical idea, it must first be constrained, and when so constrained may then be elaborated into a panoply of related ideas."

Newton: "We started with the positive integers, a subject I could not imagine held such amazing profundities. From there it was more amazement with the negative integers and then even more with quotient numbers. Quotient numbers, we discovered, harbor a topology, from which the mathematical ideas of limits germinate. From there we examined the amazing world of functions, and ideas of continuity, derivatives and integrals."

Breton: 'But none of this amazing world of Mathematics is Theoretical Physics."

Einstein: “We ended the day by showing how these mathematical ideas can be transformed into Theoretical Physics. First we would give any mathematical idea a physical label. Physical labels are all reducible to three elementary ones: for extension (L), for motion (V), and for force (F). Mathematical expressions, being unlabeled, may be combined in ways that are not allowable for Theoretical Physics. So it became apparent that, although an identical symbol might be used, a number in Mathematics is a different idea from a number in Theoretical Physics.”

Newton, continuing: "Expressions in Theoretical Physics must follow the Rules for Labels. The Rules show how new ideas for Theoretical Physics can be created from the elementary ones.”

Einstein, looking to lead: "In addition to labeling, we saw how the ideas of Theoretical Physics must be referenced either materially or locally.”

Breton: "And how the word 'set' can said of material things as well as mathematical ideas which led to the idea of a particle, the properties of material things, and the constraints of resolution.”

Newton: “All of this we put into a book enigmatically titled 'tp1.1'.”

Newton: "The next day we examined the subject of location.”

Breton: "Mathematics provides an interesting structure called *vectors* which may be suitable for transformation into Theoretical Physics. In analogy to quotient numbers, we speculated that vectors might be formed into an algebra or even into a calculus.”

Newton: "The axioms of the set of vectors allow for addition, but even addition in the set of vectors is undefined. So with some difficulty we arrived at a suitable definition for addition. Proper definition for subtraction followed easily. But multiplication was a different story.”

Einstein: “We found three different types of multiplications: inner products, cross products, and outer products.”

Newton: " We learned how to construct sums of inner products geometrically, thereby showing relationships between vectorial constructs and geometrical ones. We did the same with cross products and then the same with outer products.”

Einstein: “We climbed further to obtain the *scalar* triple product which is the inner product of one vector with the cross product of two others. But when the topic of a *vector* triple product was addressed, Breton decided to talk about the zero vector as an origin. Like any vector the origin has both a magnitude and a direction. Its magnitude is the quotient number 0. Its direction, however, may be arbitrary.”

Newton: "Assigning an arbitrary direction for the origin allowed it to be expressed in terms of a three orthogonal directions. This assignment confers on the **0** of the vector set a characteristic which is called origin. Specifying vectorial problems in terms of the origin often transforms difficult geometrical proofs into much easier bookkeeping exercises. In this way we were able to prove the vector triple product as well as a large group of vectorial identities which I summarized into an impressive table."

Einstein, resuming: "We learned how to divide vectors. Having thus constructed a vectorial algebra, we turned to solving algebraic equations. We discovered comprehensive solutions to equations with inner products, cross products and outer products."

Newton: "Then Breton introduced the idea of a matrix, explained how matrices could be added, subtracted, multiplied, and in some instances. divided. He also showed how cross products could be expressed with a matrix. So algebraic vectorial equations could be formed using matrices."

Einstein: "And with much intellectual labor we obtained comprehensive solutions for such equations."

Newton: "Which I dutifully assembled into tables for quick reference."

Einstein: "We even solved for the matrix in those cases where the matrix was the unknown."

Breton: "Finally, having completed a beautiful mathematical structure, we asked how it might be transformed into Theoretical Physics."

Newton: "First we recognized that all the vectorial multiplications have geometrical representations as areas. So in addition to scalar areas we might just as well talk about vectorial and transformational areas. Breton surprised me by showing that duals are also possible, that is material as well as local areas, and further with the assignment of physical units various combinations of extension, movement and force could be validly included. Thus an enormous inventory of ideas can be formulated for use by physicists."

Einstein: "Similarly for volumes represented by triple products."

Newton: "The discussion about references proved particularly interesting."

Breton: "The references show we can physically observe in two modes: locally or materially. Materially, when we fix our attention on the object as it moves from one location to another. Locally, when we fix our attention on one location and observe what objects pass through that location.

The experience of observation allows us to express a fundamental principle of Physics: non-collocation, namely that in a single observation no single particle can occupy two different locations. The principle of non-collocation separates Theoretical Physics from Mathematics."

Newton: "The principle constrains our vectorial algebra. Some valid mathematical vectorial operations are not permitted in Theoretical Physics, namely those which transgress the principle of non-collocation."

Breton: "After briefly reflecting on the ideas of energy and the famous formula f=ma, we ended the day's discussion by resolving to investigate if the set of vectors can be made into a calculus."

Newton: "When I reflect on our conversation yesterday I stand amazed at the many and amazing topics we wrestled with. A brief summary just omits too much. Yesterday's conversation should be made into a book!"

Breton: "Let's title it 'tp1.2'. The title would stand for theoretical physics 1.2. The 1.2 would indicate more to come."

Einstein: "Who, except us, would know what tp1.2 means?"

Breton: "We could give it a subtitle like 'A continuing inquiry into the Foundations of the Science of Physics: Vector Algebra.'"

So agreed, the friends turned to Breton to show the way to vectorial calculus.

Breton: "Our experience with vectorial algebra has shown us we often need to deal with complicated expressions which require many lines of equations. We should expect the trend to continue in our investigation of vectorial calculus. To simplify matters, I will elect, as appropriate, to unify our thinking by placing the equations in the context of a formal proof."

Straight maps in V3

Breton: "Let us start with functions with quotient numbers as the domain and the vector set as the range. For instance, for

$$\mathbf{f}(q) = \mathbf{v0} + q*\mathbf{v1}$$

where the **v**'s are constant elements of the set of vectors, and q is any quotient number. The function maps quotient numbers into a vectorial line, that is,

$$\mathbf{f}:Q \rightarrow \mathbf{V3}.$$

Newton, adding helpfully: "That's not much different from a function over Q itself, like

$$f(q) = q0 + q*q1$$

where q0, q1, and q are quotient numbers with q0 and q1 constant, and q variable."

Einstein, cautiously: "Let's see a diagram."

So Breton quickly produced the following diagram.

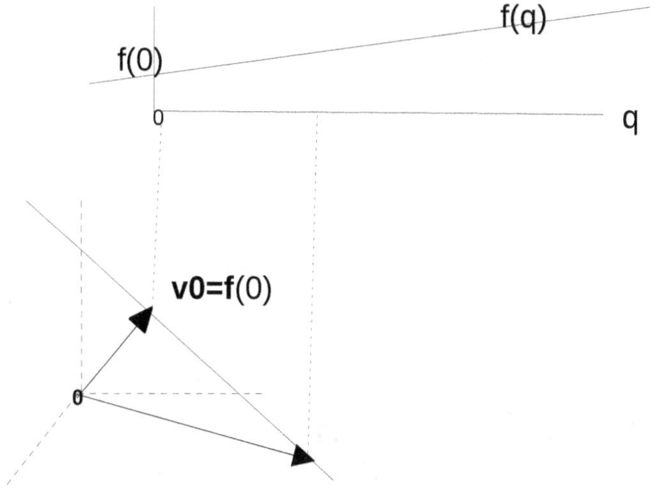

Breton: "The top diagram shows a scalar function f(q) as a line passing through f(0) at q=0 The bottom diagram shows the orientation of the origin in dashed lines and the vector function **f**(q) as a line of vectors containing **v0**. The thin dashed lines show the correspondence with the underlying field Q. The vector function may have limits, may be continuous, may have derivatives in a sense not much different that those over Q."

Newton, taking up the hint: "The derivative of f(q) is

D[q1](f(q);dq) = q1.

Breton: "Likewise the derivative of **f**(q) is

D[**v**1](**f**(q);dq) = **v1**.

Einstein, objecting: "You have not defined that derivative."

Breton, explaining: "Not in the context of the set of vectors. This derivative takes its definition from the topology of Q as defined in tp1.1. The domain of the function is Q, the range has changed from Q to **V3**. The definition of this derivative depends only on the topology of the domain. Consequently the definition of derivative given in tp1.1 applies to these functions too."

Newton, adding: "My function is a straight line in Q in the direction of q1; Breton's function is a straight line in **V3** in the direction of **v1**."

Breton: "Such a function is called a straight, linear map in **V3**."

Einstein, objecting again: "What do you mean by 'linear'?"

Breton: "You've forgotten the definition in tp1.1? Let me repeat it here."

> **Definition** (linear function)
> Given
> $f{:}X \rightarrow R$
> for
> any x1 and x2
> if
> $f(x1 + x2) = f(x1) + f(x2)$
> then
> f is a **linear** function.
> end of definition

Einstein, decisively: "So you should not call the above functions linear. Consider for

$$\mathbf{f}(q1) = \mathbf{v0} + q1{\bullet}\mathbf{v1}$$
$$\mathbf{f}(q2) = \mathbf{v0} + q2{\bullet}\mathbf{v1}$$

and

$$\mathbf{f}(q1+q2) = \mathbf{v0} + (q1+q2){\bullet}\mathbf{v1}$$

but

$$\mathbf{f}(q1) + \mathbf{f}(q2) = 2{\bullet}\mathbf{v0} + (q1+q2){\bullet}\mathbf{v1}$$

so

$$\mathbf{f}(q1+q2) \neq \mathbf{f}(q1) + \mathbf{f}(q2).$$

Breton, agreeably: "You're right. The definition would hold only for $\mathbf{v0} = \mathbf{0}$. Still the idea of lines is fundamental enough to consider enlarging our definition of 'linear'. Suppose we start by enlarging the function I first proposed to

$$\mathbf{f}(q) = \mathbf{v0} + g(q){\bullet}\mathbf{v1}$$

where g(q) is a function in Q. Then, although g(q) may not describe a line in Q, **f**(q) is nevertheless still a straight line in **V3.** So the burden of an enlarged definition of 'linear' can fall on g(q) as in the following definition."

> **Definition** (linear function)
> Given
> $\mathbf{f}{:}Q \rightarrow \mathbf{V3}$
> as a function of g(q);
> if
> g(q) is a line in Q
> then
> **f** is a **linear** function in **V3**.
> end of definition

Newton, still trying to be helpful: "This new definition contains the old one, that is, when **v0** or q0 are zero."

Breton: "With this new definition, then, my vectorial function

$$\mathbf{f(q) = v0 + q \bullet v1}$$

is a straight, linear map in V3."

Einstein, probing: "So there are straight, non-linear maps too?"

Breton: "Yes indeed, for instance,

$$\mathbf{f(q) = v0 + q^2 \bullet v1.}$$

Einstein, concluding: "So the idea of 'linear' is a subset of 'straight'!"

Breton, elucidating: "Rather they are two separate ideas. A function may be

linear and straight
non-linear and straight
linear and non-straight
non-linear and non-straight."

Newton, intrigued: "Give us an example of a linear, non-straight vectorial function."

Breton: "Sure. Consider

$$\mathbf{f(q,v) = v0 + q \bullet (v1 \wedge v)}.$$

Einstein, reflecting: "So these functions form a copious set."

Breton: "Indeed, they do. For now let us concentrate on linear, straight functions. There are yet others even in this limited category. Consider

$$\mathbf{f(q1,q2) = v0 + q1 \bullet v1 + q2 \bullet v2}$$

where **v0**, **v1**, and **v2** are constant vectors while q1 and q2 are variable quotient numbers. Such functions describe straight, linear planes in **V3**."

Newton. elaborating: "So we could also have

$$\mathbf{f(q1,q2,q3) = v0 + q1 \bullet v1 + q2 \bullet v2 + q3 \bullet v3}$$

where **v0**, **v1**, **v2**and **v3** are constant vectors while q1, q2 and q3 are variable quotient numbers. These would be vectorial volumes."

Breton: "Let us call them **regions**, rather than volumes. For each region we could calculate a volume, but the relevant equation itself describes only the region itself."

Newton: "Agreed. You said these functions have derivatives."

Breton, reflecting: "Perhaps too hastily. In the equation

$$\mathbf{f}(q) = \mathbf{v0} + g(q)\bullet\mathbf{v1}$$

the function g(q) may not have a derivative, or if it has a derivative, it need not be basic. If the function has basic derivative, only then does

$$D[q1](\mathbf{f}(q);dq) = D[q1](g(q);dq)\bullet\mathbf{v1}$$

where q1 may be variable.

The unique and constant direction, **uv1**, is called the **straight line's direction**. The constant vector **v1**, is called the line's **direction vector**. Do you remember our discussion about restricted functions and restricted topologies?"

Einstein, dissembling: "Yes."

Breton, continuing: "As a function of **V3**, the function can be considered a restricted function with a restricted topology."

Newton, dampening the enthusiasm with a cautious note: "What is the topology of **V3**?"

The topology of V3

Breton, enthusiastically: "Consider the set VT consisting of subsets of **V3** which are arbitrary unions and intersections of sets

$$N = \{\mathbf{v} | abs(\mathbf{v}-\mathbf{v1}) < eq\}$$

for all **v1** of **V3** and all positive eq of Q. Any $Ni = \{\mathbf{v} | abs(\mathbf{v}-\mathbf{vi}) < eqi\}$ is called an open set of VT. Is VT is a topology."

Newton, candidly: "Refresh my memory about a topology."

Breton: "You can look it up in tp1.1 page 239. These are the requirements:

- **V3** is a member of VT
- The null set is a member of VT
- If the subsets N1 and N2 are members of VT,
  then N1 ∩ N2 is also a member of VT (the intersection property)
- If the subsets N1 and N2 are members of VT,
  then N1 ∪ N2 is also a member of VT (union property)
- The intersection property holds for a finite number of intersections
- The union property for an infinite number of unions."

Einstein, entering the discussion: "So let us see if VT is a topology. Is **V3** in VT?"

Breton, methodically: "Consider the infinite union of all $N = \{\mathbf{v} | abs(\mathbf{v}) < eq\}$ for every possible eq. Such a union includes every member of **V3**, and so also **V3**."

Einstein, continuing: " I can see that the null set is in VT since the intersection of two disjoint subsets is null and such subsets abound in VT."

Newton, joining in: "How about the intersection of any two members of VT?"

Breton: "Both the intersection and union properties are included in the definition of VT."

Newton, continuing specifically: "How about finite intersections?"

Breton: "Let

$$N1 \cap N2 \cap N3 \ldots\ldots\ldots\ldots \cap Nn$$

be any finite intersection of subsets of VT.
If any of the intersections is null then the finite intersection is also null, and so a member of VT. If not, then the intersection is an open subset of VT. Let **vi** be a member of the finite intersection. Then for some eq

$$Ni= \{\mathbf{v}|abs(\mathbf{v-vi})<eq\}$$

is a subset of the intersection. Consequently the finite intersection consists of an arbitrary union of such sets. This union is a member of VT. We can conclude then that any finite intersection of subsets of VT is also is a member of VT."

Newton, continuing finally: "How about the union property?"

Breton: "We have to appreciate that VT consists of a great many subsets. If the infinite union is one of disjoint subsets, then their arbitrary union is included by the definition of VT. So consider the infinite union of subsets which are not disjoint. If the infinite union is V3, then it is included in VT. If not, then the infinite union is a the union of an arbitrary number of subsets of VT which is also included in VT. We can conclude then that any infinite union of subsets of VT is also is a member of VT."

Newton, concluding: "So **V3** does indeed harbor a topology."

Breton, continuing: "Do you remember the definition of a restricted topology?"

Newton: "Please help my memory."

Breton: "Let me recast the former definition specifically for our set of vectors."

Definition (restricted topology)
Given
VT, the topology of **V3**;
S, any open set of VT
U, an arbitrary subset of **V3**
then
$$VT|U \equiv \{S \cap U\}$$
is called **the topology of V3 restricted to U**.
end of definition

Einstein: "What good is this definition?"

Breton: "It will serve when we consider lines and planes and other limited subsets of **V3**. The subsets of U may also harbor a topology, different from but related to VT."

Newton, questioning: "So we can have open sets in lines?"

Breton: "Not as such. The sets of VT|U are the open sets of U. The open sets of U may contain sets which are *not* open in VT."

Einstein, with a touch of impatience: "We have a topology. Move on to the limits in **V3**!"

Breton happily continuing: "Since **V3** harbors a topology, we can discuss limits in **V3**. First permit me to define a few other useful ideas.

Let VM be the smallest set of subsets of **V3** which includes all its open sets, their complements, and all denumerable unions of such sets. The sets of VM are called the **measurable** sets of **V3**."

Einstein: "Won't we meed a metric?"

Breton: "Yes, of course. Since
$$0 \leq \text{abs}(\mathbf{v}-\mathbf{v1}) \leq \infty$$
$$0 = \text{abs}(\mathbf{v}-\mathbf{v1}) \text{ if and only if } \mathbf{v} = \mathbf{v1}$$
$$\text{abs}(\mathbf{v}-\mathbf{v1}) = \text{abs}(\mathbf{v1}-\mathbf{v})$$
$$\text{abs}(\mathbf{v}-\mathbf{v1}) \leq \text{abs}(\mathbf{v}-\mathbf{v2}) + \text{abs}(\mathbf{v2}-\mathbf{v1}),$$
magnitude may serve as a metric."

Newton, probing: "In Q we have limits from above and from below. In **V3** we have not only two directions, but an infinite number."

Breton: "We have first to discuss limit vectors."

Newton: "Remind me about limit vectors."

Breton: "The definition given in tp1.1 will do. Here it is applied to **V3**."

Definition (limit vectors)

Given

VT the topology of **V3**

v1, a vector in **V3**

Q the set of quotient numbers

for

eq, any positive number of Q

then

v1 is a **limit vector** of **V3** if every open set $\{\mathbf{v}|\text{abs}(\mathbf{v}-\mathbf{v1})<\text{eq}\}$ contains a vector different from **v1**.

end of definition

Einstein: "Illustrate what you are saying."

Breton: "This illustration may help."

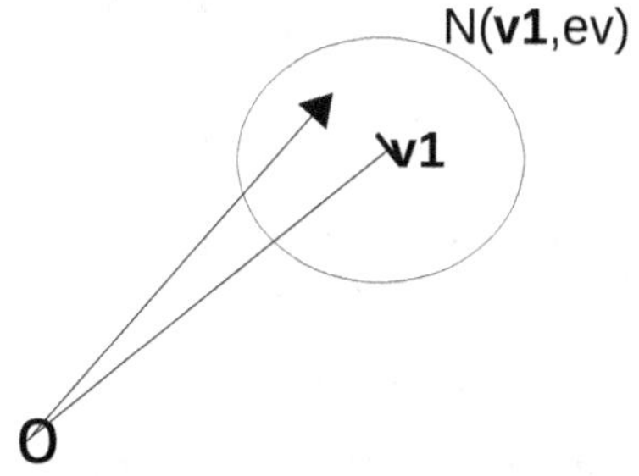

Breton: "Imagine the circle to be a spherical open set about **v1**. If **v1** is a limit vector, then it will always be possible to find another vector in the open set which is different from **v1**, no matter how closely the open set encircles **v1**."

Newton, questioning for clarity: "Show us an illustration of a vector which is not a limit vector."

Breton then produced the following illustration.

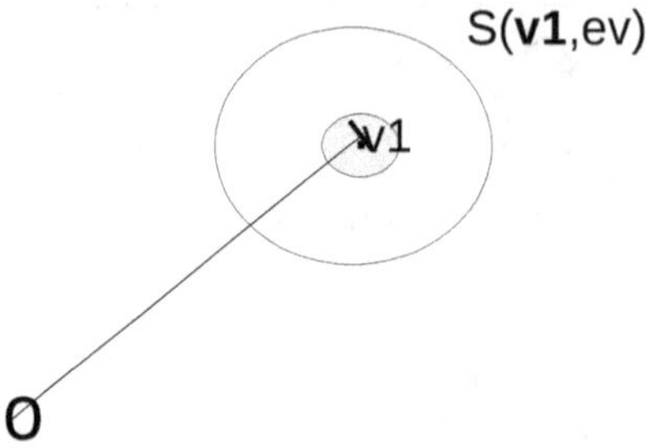

Breton: "Imagine **v1** in the center of a doughnut with no vectors other than itself in the doughnut hole."

Newton: "So limit vectors are vectors which enjoy a certain affinity to the vectors which surround them."

Einstein. concluding: "Then every vector in **V3** is a limit vector. But a corresponding statement may not generally be made of VT|U."

Newton, questioning: "So what is a limit in **V3** with its infinite directions?"

Limits in V3

Breton: "We will have to consider more than one possibility. Let's start by considering an inner product like f(**v**) = **v**•**q**(**v**)."

Newton: "That function equals the quotient number 3, for any **v.**"

Breton: "For this and similar functions a limit may be specified directly from, the topology of VT."

Definition (omni-directional limits in **V3**)
Given
ef and ev, positive quotient numbers;
v1 and the open set {**v**|abs(**v**−**v1**)<ev};
f(**v**);
for
dv = **v**−**v1**, **v** any vector in the open set;
if there can be found a vector **f1** such that
for any ef >0 there can always be found an ev> 0 such that
abs(**f**(**v1**+**dv**) − **f1**)<ef
for any **dv** in the open set
then
f1 is **the omni-directional limit** of **f** at **v1**.

end of definition

Newton, adding helpfully: "A similar definition can be made for scalar functions."

Breton: "Of course. Such a limit, if it exists, is written

lim(**f**(**v1**+**dv**)) = **f1** as **dv**→**0**

or lim(**f**(**v1**+**dv**)) → **f1** as **dv**→**0**.

Please note that **f1** need not equal **f**(**v1**)."

Einstein: "What do you call functions for which **f**(**v1**) = **f1**?"

Breton: "Such functions are called **omni-directionally continuous** at **v1**. For such functions,

f(**v1**) = lim(**f**(**v1**+**dv**)) as **dv**→**0**.

Einstein, happily leading: "It strikes me that there can exist functions which do not have limits in the omni-directional sense, but still have limits in some secondary sense."

Breton: "Yes indeed. You have opened the way to a large subject. Let us start with directional limits."

Einstein: "We need a definition."

Breton: "Of course. Here it is."

Definition (directional limits in **V3**)

Given

ef and dq, positive quotient numbers;

v, **v1** vectors in **V3**;

uv ≡ **u**(**v**−**v1**), a direction;

f(**v**), a given vector function;

if there can be found a vector **f1** such that

for any ef >0 there can always be found an dq > 0 such that

abs(**f**(**v1**+dq****uv**) − **f1**)<ef

then

f1 is **the forward uv directional limit** of **f** at **v1**.

if there can be found a vector **f1** such that

for any ef >0 there can always be found an dq > 0 such that

abs(**f1** − **f**(**v1**−dq****uv**))<ef

then

f1 is **the backward uv directional limit** of **f** at **v1**.

end of definition

Newton, questioning for clarity: "Would the definition also work for scalar functions, f(**v**)?"

Breton: "Yes, indeed.
Directional limits are symbolized as

$$\lim_f \mathbf{f}|\mathbf{uv} = \mathbf{f1} \text{ at } \mathbf{v1} \text{ as } dq \to 0$$
$$\lim_f f|\mathbf{uv} = f1 \text{ at } \mathbf{v1} \text{ as } dq \to 0$$
$$\lim_b \mathbf{f}|\mathbf{uv} = \mathbf{f1} \text{ at } \mathbf{v1} \text{ as } dq \to 0$$
$$\lim_b f|\mathbf{uv} = f1 \text{ at } \mathbf{v1} \text{ as } dq \to 0$$

or

$$\lim(\mathbf{f}(\mathbf{v1}+dq*\mathbf{uv})) \to \mathbf{f1} \text{ as } dq \to 0$$
$$\lim(f(\mathbf{v1}+dq*\mathbf{uv})) \to f1 \text{ as } dq \to 0$$
$$\lim(\mathbf{f}(\mathbf{v1}-dq*\mathbf{uv})) \to \mathbf{f1} \text{ as } dq \to 0$$
$$\lim(f(\mathbf{v1}-dq*\mathbf{uv})) \to f1 \text{ as } dq \to 0.$$

Newton: "These directional limits look like the limits in Q with their forward and backward definitions."

Breton: "They can be viewed as limits of restricted functions f((**v**)| dq***uv** in the topology restricted to the **uv** direction. Please note that at **v1** the directional limit in one direction **uv1** may differ from one in another direction **uv2**."

Einstein, objecting: "But these limits may not exist at all."

Breton, agreeing: "Yes, and further the forward directional limit may not equal the backward limit, nor need either equal **f(v1)**. A function is said to be **continuous at v1 in the uv direction** if

$$\mathbf{f}(\mathbf{v1}) = \lim(\mathbf{f}(\mathbf{v1}+dq*\mathbf{uv})) = \lim(\mathbf{f}(\mathbf{v1}-dq*\mathbf{uv})) \text{ as } dq \to 0.$$

Einstein, enjoying setting the agenda: "Are there other classes of limits which still preserve a sense of direction but still apply to a larger class of functions that those with a simple omni-directional limit?"

Breton: "A good question whose answer is 'yes'. The arbitrary orientation of the set of vectors provides just such a possibility. In this classification an arbitrary vector

$$\mathbf{v} = v* (c1 \bullet \mathbf{u1} + c2 \bullet \mathbf{u2} + c3 \bullet \mathbf{u3})$$

is classified by the sign of its directions, c1, c2, and c3. There are 8 such gross "directions" which are called **quadrants**. The quadrant for which the ci are all non-negative is called the **positive quadrant**."

Einstein: "Show us an illustration of the eight quadrants."

Breton: "Gladly. The darkened lines show the first quadrant; the dashed lines are the corresponding negative directions. You can count the eight different quadrants by selecting three different directions at a time."

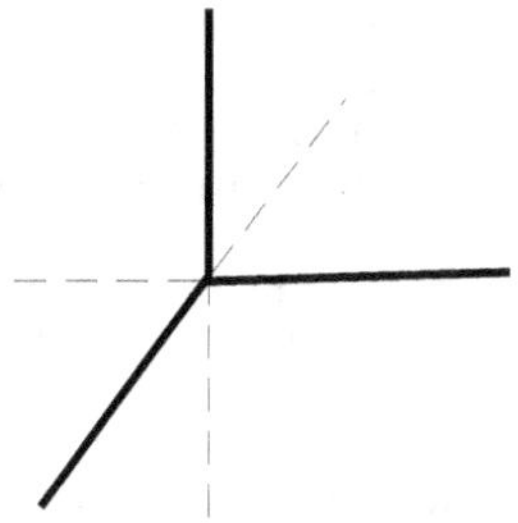

The eight quadrants

Einstein: "So define these limits!"

Breton: "Let me start with the positive quadrant."

Definition (limits *of* the positive quadrant)
Given
dq1, dq2, dq3 positive quotient numbers,
f(**v**), a given vector function,
if there can be found a vector **f1** such that
lim((**f**(**v1**+dq1*u1)) = **f1** as dq1⟶0
and
lim((**f**(**v1**+dq2*u2)) = **f1** as dq2⟶0
and
lim((**f**(**v1**+dq3*u3)) = **f1** as dq3⟶0
and
then
f1 is the **limit of f of the positive quadrant at v1**.

end of definition

Newton, adding as usual: "The definition applies to scalar functions too."

Breton: "Yes, of course. Limits of the positive quadrant can be seen as limits in the topology VT restricted to the directions **u1**, **u2**, and **u3**."

Newton: "How do you symbolize these limits?"

Breton: "Another good question. Since these limits differ from the others, they need a separate symbology. Here it is.
lim((**f**(**v1**+{**dv**}) = **f1** as **dv+⟶0**
or
f(**v1**+{**dv**}) ⟶ **f1** as **dv+⟶0**

Einstein, inquisitively: "So the quadrant limits are just directional limits in specified directions."

Breton: "Not really. A function which had a different directional limit in each of the specified directions would not be a limit of the first quadrant. Functions with limits of the positive quadrant have the *same* limit in the three direction of the origin."

Newton, reflecting: "I'm beginning to appreciate more and more how numerous are the mathematical objects symbolized by the word 'function.'"

Breton: "The study of Mathematics requires not only a logical mind, but also an imaginative one. Can you imagine functions which have the same limit for any direction in the positive quadrant?"

Einstein, taking the hint: "Define those limits please."

Breton: "Here it is."

Definition (limits *in* the positive quadrant)
Given
uv, a direction in the positive quadrant;
f(v), a given vector function over **V3**;
dq, a positive quotient number;
if there can be found a vector **f1** for all **uv** in the positive quadrant such that

lim((**f(v1**+dq***uv**)) = **f1** as dq→0

then
f1 is the **limit of f in the positive quadrant at v1**.

end of definition

Newton, adding as usual: "The definition applies to scalar functions too."

Breton: "Yes, of course. Limits in the positive quadrant can be seen as limits in the topology VT restricted to the positive quadrant."

Newton: "How do you symbolize these limits?"

Breton: "Since these limits differ from others, they too need a separate symbology. Here it is.

lim((**f(v1+dv**) = **f1** as **dv+→0**

or

f(v1+dv) → **f1** as **dv+→0.**

Newton: "So that is why the {}'s were used to symbolize limits *of* the quadrant, so they could be contrasted to limits *in* the quadrant. The {**dv**} is used for the set of three directions, whereas the **dv** is used for any direction."

Einstein, reflecting: "If a function has a limit *in* the first quadrant, then it must also have a limit *of* the first quadrant."

Breton: "True enough, but the reverse is not true generally."

Newton, reflecting: "Limits in the other quadrants could also be defined, and such limits may not equal the limit of the first quadrant. Breton, you have stretched my imagination beyond imagining."

Breton: "There is still much more. More stretching is needed. The idea of the positive quadrant can be broadened to an arbitrary section."

Einstein: "What is a section?"

Breton: "It deserves a definition."

Definition (section)
Given
sqi, i=1,2,3 positive quotient variables;
uiV1, i=1,2,3 non parallel directions;
for **v1** as a reference
the set
SECT(**v1**,**u1V1**,**u2V1**,**u3V1**)
≡ {**v**|**v** = **v1**+ sq1*****u1V1** + sq2*****u2V1** + sq3*****u3V1**}
for all sqi is called a **section of V3 at v1**.

The subsets
{**v**|**v** = **v1**+ sq1*****u1V1** + sq2*****u2V1**}
{**v**|**v** = **v1**+ sq1*****u2V1** + sq3*****u3V1**}
{**v**|**v** = **v1**+ sq2*****u2V1** + sq3*****u3V1**}
are called the **faces** of the section.

end of definition

Einstein, objecting: "Aren't you confusing things? Suppose we have a given vector **v2** in a section. First you say that **v2** has a representation in the arbitrary orientation as

v2 = **v1** + q1*****u1** + q2*****u2** + q3*****u3**

for some q1, q2, q3
but also in the sector as

v2 = **v1** + sq1*****u1V1** + sq2*****u2V1** + sq3*****u3V1.**

Breton: "Very little gets by you, Einstein. Both representations are valid, so we have two ways of referring to a given vector. There are differences, however. For vectors represented sectionally the sqi are positive while the **uiV1** may have any direction. So the sectional representation always implies the positive-definite convention. In contrast, the representation using the origin the qi may be negative but the **ui** are positive in the given orthogonal orientation. This is the basic convention."

Newton, commenting: "Positive sqi means that the vertex of a section excludes reflex angles. I'm curious. Is there a way to transform these sections?"

Breton: "Yes. To transform vectors oriented to the origin, we can use the solution found in tp1.2. If the vector **v1** = q11***u1** + q12***u2** + q13***u3** is to be transformed into the vector **v2** =q21***u1** + q22***u2** + q23***u3** we can use the matrix **qd(v1)*v2** since **v1 • qd(v1)*v2** = **v2.** This transformation transforms the q1i's into the q2i's. Something similar can be done for the sectional representation, but I ask for your patience until the idea of a section is developed further."

Einstein, leading: "I suppose we can define limits for sections."

Breton: "Yes, you recognize the way. First this definition."

Definition (limits *of* section)
Given
- **v1** as a reference;
- SECT(**v1**,**u1V1**,**u2V1**,**u3V1**);
- **f(v)**, a given vector function;
- sqi, i=1,2,3 positive quotient variables;
- **uiV1**, i=1,2,3 non parallel directions;

if there can be found a vector **f1** such that

lim((**f(v1**+sq1***u1V1**)) = **f1** as **sq**1⟶0

and

lim((**f(v1**+**s**q2***u2V1**)) = **f1** as **s**q2⟶0

and

lim((**f(v1**+**s**q3***u3V1**)) = **f1** as **s**q3⟶0

then

f1 is the **limit of f in SECT(v1,u1V1,u2V1,u3V1) at v1**.

end of definition

These limits are symbolized as

lim((**f(v1**+{**dV1**}) = **f1** as **dV1**⟶0

or

f(v1+{**dV1**}) ⟶ **f1** as **dV1**⟶0.

Then this one

> **Definition** (limits *in* section)
> Given
> **v1** as a reference;
> SECT(**v1**,**u1V1**,**u2V1**,**u3V1**);
> **uv**, a direction in SECT(**v1**,**u1V1**,**u2V1**,**u3V1**);
> **f**(**v**), a given vector function;
> sq, a positive quotient number;
> if there can be found a vector **f1** for any **uv** such that
> lim((**f**(**v1**+sq*****uv**)) = **f1** as sq→0
> then
> **f1** is the **limit of f in SECT(v1**,**u1V1**,**u2V1**,**u3V1**) **at v1**.
>
> end of definition

Limits in section are symbolized as
lim((**f**(**v1**+**dV1**) = **f1** as **dV1**→0
or
f(**v1**+**dV1**) → **f1** as **dV1**→0
Limits in section may be viewed as limits in a topology restricted to the section."

Newton, reflecting: "Every function with an omni-directional limit at **v1** also has a sectional limit there."

Breton: "Sectional limits are exemplified in quadrant limits on the one hand, and as directional limits on the other. Consequently they occupy a central consideration in a discussion about limits in **V3**."

Newton, reflecting: "We are moving towards more and more restricted functions. In the set of vectorial functions there must exist subsets of more narrowly restricted functions."

Breton: "Some of which are proper subsets of subsets of less restricted functions. We would do well to keep this hierarchy in mind since misplacing a function in the wrong subset will inevitably lead to confusion and error."

Einstein, looking again to lead: "Are there no limits in vectorial curves?"

Limits along curves

Breton, accepting the change in direction: "Why yes. A curve serving as a restricted set may also form the basis for functional limits. But what is a curve?"

Einstein, disdainfully: "Everyone can recognize a curve."

Breton: "But an unrestricted collection of vectors {**v**} does not describe a curve."

Einstein: "Then can *you* define a curve?"

Breton: "First let me define when two vectors are connected."

Definition (connection)
Given
eq > 0, a positive quotient number;
N(**v1**,eq), an open neighborhood of **v1;**
N(**v2**,eq), an open neighborhood of **v2**;
if for any eq
N(**v1**,eq)∩N(**v2**,eq) is not empty
then
v1 is connected to v2
end of definition

Now we can define a continuous curve."

Definition (continuous curves and surfaces)
Given
S, a subset of **V3** consisting only of connected vectors;
eq > 0, a positive quotient number;
N(**v**,eq), an open neighborhood of **v**;
for
ut(v), a direction perhaps varying for each **v**,
if for every **v** in **S**
N(**v**,eq)∩S = {**v** + dq***ut(v)**,abs(dq) < eq} as eq⟶0
then
S is called **a continuous vectorial curve**;
for
ut1(v) and **ut2(v)**, separate directions,
if for every **v** in **S**
N(**v**,eq)∩S = {dv1***ut1(v)** + dv2***ut2(v)**,
abs(dq1) < eq and abs(dq2) < eq}
as eq⟶0
then
S is called **a continuous vectorial surface**.
end of definition

Newton: "How do you deal with segments of a curve?"

Breton: "The continuity of **S** arises from its connectedness. If **S1** and **S2** are each connected curves with **S1∩S2** empty, then **S1∪S2** is said to be curve composed to **two disconnected segments**."

Einstein, inquiringly: “There must be some connection to the straight maps we discussed earlier.”

Breton: "Some curves can also be described with reference to the underlying field Q. Consider **v**(q) differentiable with respect to q. Then since the derivative D[x1](**v**(q);dq) is a vector with a connected direction

CQ ≡ {**v**(q)}

is a continuous curve.”

Newton, somewhat perplexed: "Give us an example!”

Breton: "Of course. Can you see what this curve looks like?

v(q) = q•**v1** + sin(q)•**v2** + cos(q)•**v3**

for **v1**, **v2**, **v3** fixed.”

Newton, after a few minutes doodling with a pencil and a pad: "It is and elliptical helix in **V3**. Its axis is **v1**; the ellipticity is defined by **v2** and **v3**.”

Einstein, with a note of impatience: "So we can describe complicated curves in **V3** this way. Can our earlier description of linear curves be modified to include these complicated curves?”

Breton: "Earlier we began with **f**(q) = **v0** + q•**v1**. We saw we could possibly have

linear and straight
non-linear and straight
linear and non-straight
non-linear and non-straight

functions.”

Newton: "Let me construct a table to symbolize these functions.”

Category	Symbol	Example
linear, straight	**f**(q)	**v0** + q∗**v1**
non-linear, straight	**f**(q)	**v0** + g(q)∗**v1**
linear, non-straight	**f**(q,**v**)	**v0** + q∗**g**(**v**)
non-linear, non-straight	**f**(q,**v**)	**v0** + g1(q)∗**g2**(**v**)

Breton: " Thank you. The table is helpful. Let me develop this a little. Consider

f(q1,q2) = **v0** + q1•**v1** + q2•**v2**

and **f**(q1,q2,q3) = **v0** + q1•**v1** + q2•**v2** + q3•**v3**

where **v0**, **v1**, **v2**, and **v3** are constant, non-parallel vectors and q1, q2, and q3 are variables.

The function **f**(q1,q2) describes a linear surface (plane); the function **f**(q1,q2,q3) describes a linear vectorial vicinity. Such functions are also called straight maps."

Newton: "And we can also have straight, non-linear functions as we showed earlier."

Einstein: "How about linear, but non-straight?"

Breton: "These hold the q's constant but allow variable vectors. For illustration

f(**v**) = **g0**(**v**) + q1•**g1**(**v**)

with q1 fixed."

Newton: "So we could also have functions like

f(**v1**,**v2**) = q1•**g1**(**v1**) + q2•**g2**(**v2**)

and **f**(**v1**,**v2**,**v3**) = q1•**g1**(**v1**) + q2•**g2**(**v2**) + q3•**g3**(**v3**)

with the qi's fixed and the **vi**'s variable."

Einstein, conclusively: "So we could have **f**(q,**v**) for functions which are neither straight nor linear."

Newton, adding: "And up to three dimensions."

Breton, concluding: "So we have created a large category of functions which might happily be used for describing many smooth physical observations."

Newton: "What category is our elliptical helix?"

Breton: "It has fixed **vi**'s and variable, non-linear qi's. So it is a non-linear, straight function."

Newton: "It is a curve and not a surface or a region."

Breton: "Because the helix is a function of only one variable, albeit a non-linear one."

Einstein: "What more would you like to say about these functions?"

Breton: "A few more definitions focused on one dimensional straight functions, **f**(q). At each vector on the curve let me define a tangent line as

tang(q) ≡ **f**(q1) + q•D[q1](**f**(q);dq)

where **f**(q1) is the reference. The direction of the tangent line is the direction of D[q1](**f**(q);dq)."

Einstein: "So give us an example using the elliptical helix."

Breton: "For the elliptical helix

D[q1](**v**(q);dq) = **v1** + cos(q1)•**v2** – sin(q1)•**v3** so the tangent line at **v**(q1) is

tang(q) = (q1+ q)•**v1**
+ (sin(q1)+cos(q1))•**v2**
+ (cos(q1)–q*sin(q1))•**v3**

Einstein: "Anything else?"

Breton: "Now we can discuss arc-lengths."
I“
Einstein: "What is an arc-length?"

Breton: "The arc-length of a curve, **f**(q), referenced to **f**(q1) is defined as

s(**f**(q1),**f**(q)) ≡ I[q1,q](abs(D[t](**f**(u);du));dt).

Arc-lengths increase monotonically."

Einstein: "Why bring up these subjects now?"

Breton: "Because they help in the analysis of curves."

Einstein, leading by objecting: "But they divert us now from limits and derivatives. Let's put arc-lengths aside for now. Continue with limits."

Newton: "Yes, are you done with limits?"

Breton: "Well no. We need to discuss limits along these curves."

Einstein: "Then do it!"

Breton: "Functions over q may be either restricted, **f**(q)|**v**(q), or compound, **f**(**v**(q)). First let us start with restricted functions. Ideas derived from quotient functions (such as continuity, step functions, derivatives, and integrals) can be extended into the subset {**v**(q)}. As maps of quotient numbers into the set of vectors, these restricted functions possess the same properties (and restrictions) as vectors themselves and other additional properties that derive from their status as functions."

Einstein: "Too many words. A definition would be preferable."

Breton: "As you prefer."

Definition (limits of restricted functions along the curve CQ)

Given

CQ = {**v**(q)}, a curve in **V3**;
q1, a quotient number;
ef and dq, positive quotient numbers;
v and **v1** = **v**(q1), vectors in **V3**;
f(q)|CQ, a given restricted function;

if there can be found a vector **f1** such that
for any ef>0 there can always be found a dq>0 whenever
abs(**f**(q1+dq)|CQ – **f1**)< ef}

then
f1 is **the forward limit along** CQ of **f** at **v**(q1);

if there can be found a vector **f0** such that
for any ef>0 there can always be found a dq>0 whenever
abs(**f**(q1–dq)|CQ – **f1**)< ef}

then
f0 is **the backward limit along** CQ of **f** at **v**(q1).;

end of definition

Limits of restricted functions along CQ are symbolized as

$\lim_f$ **f**|CQ = **f1** as dq→ 0
$\lim_b$ **f**|CQ = **f0** as dq→ 0

or

f(q1+dq)|CQ → **f1** as dq→ 0
f(q1–dq)|CQ → **f0** as dq→ 0.

Newton, as usual: "The definition also works for scalar functions."

Breton: "These limits need not exist, the function's forward limit need not equal its backward limit, not need either equal **f**(**v**(q1))."

Newton: "Just like quotient functions themselves."

Breton: "A function is called continuous in q at **v**(q1) if
f(**v**(q1))|CQ = lim **f**(**v**(q1+dq))|CQ = lim **f**(**v**(q1–dq))|CQ
as dq→ 0.

Newton: "Are there other limits?"

Breton: "Limits for curves may be defined directly from **V3** without reference to Q. Such curves are designated CV. These curves may also be the domain of compound scalar functions f(**v**(q))."

Definition (limits of compound functions along a curve CV)

Given

CV, a curve in **V3**;
v1, a vector in CV;
ut(v1), the curve's direction at **v1**;
ef and dq, positive quotient numbers,
f(v)|CV, a vector function over CV';

for

v, another vector on the curve;
dv(v1) ≡ dv•**ut(v1)**, dv>0;

if there can be found a vector **f1** such that

for any ef>0 there can always be found a dv>0 whenever
abs(**f(v1**+**dv(v1))**– **f1**)< ef}

then

f1 is a **limit of f(v) at v1 along** CV.

end of definition

Limits of functions along CV are symbolized as

$\lim_f$(**v1**+**dv(v1)**) = **f1** at **v1** as dv⟶ 0

or

f(v1+**dv(v1)**|CV ⟶ **f1** at **v1** as dv⟶ 0

Einstein, injecting: "Such curves may have more than one direction."

Breton: "Correct. And for that reason more than one limit of **f(v)** may exist there. Such functions are said to be continuous in **v** at **v1** if

f(v1) = lim **f(v1**+**dv(v1)** as dv⟶ 0 for all possible **dv(v1)**."

Newton: "These are curious functions. They may have many different values simultaneously."

Breton: "A restricted function, **f**(q)|CQ, may have a single value over an interval of q while a compound function, **f(v**(q))|CV, may have multiple values at **v**(q1). In the former case, the single functional value is called a **stalled** value; in the latter case the multiple functional values are called **branched** values."

Newton: "Some illustrations would help."

Breton: "I can only illustrate. Perhaps these might be helpful."

With that Breton sketched the following illustrations which he handed to his friends.

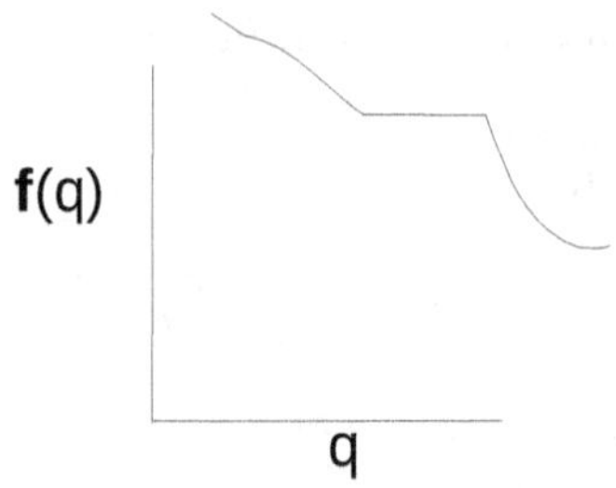

Stalled curve

Breton: " The first illustrates a stalled curve. Remember the variable q is ordered. The illustration show the three dimensional **f**(q) as a single axis. With increasing q the curve reaches a stalled value during which it does not change value, even though q continues to increase. With increasing q the curve may become unstalled."

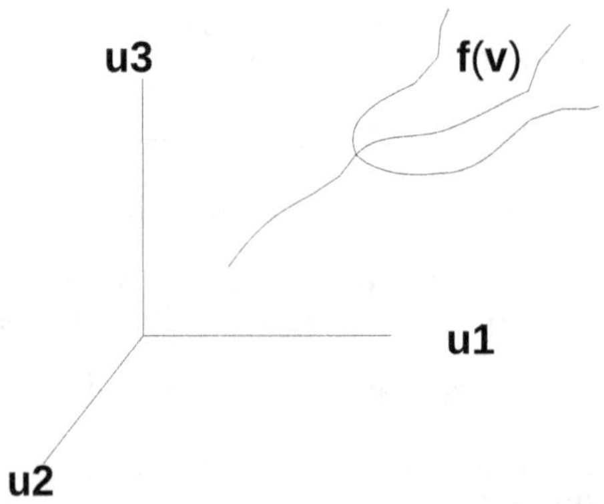

Curve with branches

This second curve illustrates a curve with branches. No order is apparent. At a certain value of the function it branches into several paths."

Einstein: "Directional limits, then, are specific instances of limits along CQ or CV."

Breton: "Correct."

Newton: "Let me put the symbolism for these many limits into a table."

Type of Limit	Symbol
omni-directional	**f(v1 + dv)** ⟶ **f1** as **dv** ⟶ **0**
directional	**f(v1 + dv)**|**uv** ⟶ **f1** as dv ⟶ 0
first quadrant	
of quadrant	**f(v1 + {dv})** ⟶ **f1** as **dv+** ⟶ 0
in quadrant	**f(v1 + dv)** ⟶ **f1** as **dv+** ⟶ 0
sectional	
of section	**f(v1+dV1**}) ⟶ **f1** as **dV1** ⟶ 0
in section	**f(v1+dV1)** ⟶ **f1** as **dV1** ⟶ 0
along a curve	
restricted	f(q1 + dq)|CQ ⟶ **f1** as dq ⟶ **0**
compound	**f(v1+dv(v1))**|CV⟶**f1** as **dv**⟶**0**

Differentiation in V3

Einstein, concluding: "Each of the limits can be used to define derivatives in **V3**."

Breton, continuing: "In **V3** differentiation may be defined with reference to the underlying field Q or to **V3** itself. With respect to Q the differentiation maintains order in the usually restricted topology. For this reason it is often referred to as **process**. The unrestricted topology, in contrast, fails to maintain order under its metric."

Newton: "Be more explicit."

Breton: "Differentiation by means of scalar increments produces three derivatives:

in a given direction and thus called directional,
along CQ with respect to q,
along CV with respect to **v**.

Differentiation by means of vector increments arises from three possibilities:

V3⟶Q called the **divergence**, symbolized by D•
V3⟶**V3** called the **curl**, symbolized by D∧
Q⟶**V3** called the **gradient**, symbolized by D✱

Derivatives in **V3**, like those in Q, belong to functions rather than to relationships and do not necessarily arise from functional values. A smaller class of functions, called **basic**, possess derivatives from which an accurate estimate of neighboring functional values may be made."

Einstein: "A large vista opens before us. Start exploring."

Directional Derivative

Breton: "From directional limits directional derivatives may be defined. But first I need to define a convention which is illustrated in the following diagram.

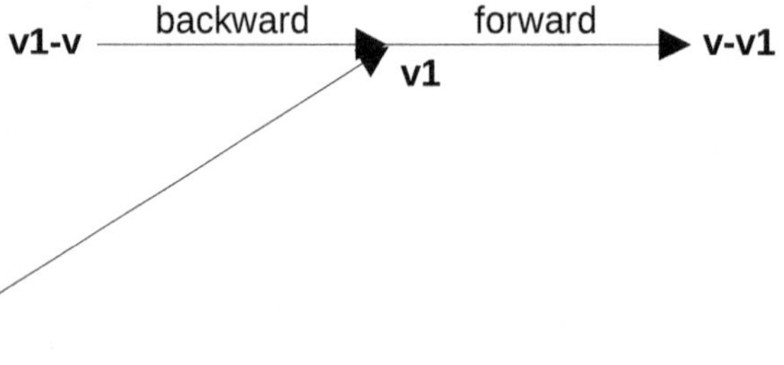

The forward sense is designated from **v1** forward (away from **v1**); the backward sense is designated toward **v1**.

Newton: "Why not both away from **v1**?

Breton: "Suppose a continuous derivative at **v1**. Then we would like to say that the backward and forward derivatives are the equal. The illustration supports such declarations. Otherwise, the backward derivative would have to be expressed as the negative of the forward one. The convention illustrated avoids such awkwardness."

Einstein: "So proceed to the definition."

Breton: "Here it is.

Definition (basic directional derivatives)
Given
v1, a reference vector in **V3;**
uv ≡ **u**(**v**−**v1**), a direction;
f(**v**), a given vector function;
then if it exists

lim(**f**(**v1**+dq∗**uv**) − **f**(**v1**))/dq as dq ⟶ 0

is called the **basic forward directional derivative of f in the uv direction**;

then also if it exists

lim(**f**(**v1**) − **f**(**v1**−dq∗**uv**))/dq as dq ⟶ 0

is called the **basic backward directional derivative of f in the uv direction**;

end of definition

Forward directional derivatives are symbolized as

D[**v1**,**v1**+dq∗**uv**](**f**(**v**)|**uv**;dq)

Backward directional derivatives are symbolized as

D[**v1**−dq∗**uv**,**v1**](**f**(**v**)|**uv**;dq)."

Newton, looking to save a little face: "So then negative derivatives may be similarly defined as

D[**v1**,**v1**+dq∗**uv**](**f**(**v**)|−**uv**;dq)
D[**v1**−dq∗**uv**,**v1**](**f**(**v**)|−**uv**;dq)."

Breton: "Yes. but be careful with the definitions.
If a function has a basic derivative at **v1**, it is call **differentiable** there. If the forward and backward derivatives are equal there, the function is said to have a **continuous directional derivative** there. If a function has a continuous directional derivative at each point on the line {**v1** + q∗**uv**}, the function is said to have a continuous derivative **everywhere** along the line."

Newton: "Can't the symbolism be simplified for continuous derivatives?"

Breton: "Good idea. Let us write continuous derivatives simply as
D[**v1**](**f**(**v**)|**uv**;dq)."

Einstein, leading again: "If the function has a basic derivative, how does its relate to other close values."

Breton: "Then,

f(**v1**+dq***uv**) ≈ **f**(**v1**) + dq*D[**v1**,**v1**+dq***uv**](**f**(**v**)|**uv**;dv))

As we saw before, this equation would hold in the limit for any D[**v1**,**v1**+dq***uv**](**f**(**v**)|**uv**;dv)) since

f(**v1**+0***uv**) = **f**(**v1**) + 0*D[**v1**,**v1**+dq***uv**](**f**(**v**)|**uv**;dv)).

Newton reflecting: "The set of functions with basic derivatives is a much smaller set than those with derivatives. For most functions with derivatives we may not use that approximation so useful to my illustrious ancestor."

Breton, acknowledging: "Yes, your illustrious ancestor would have been mightily helped by a knowledge of Theoretical Physics."

Einstein, impatiently: "Move on to other derivatives!"

Breton: "Consider next derivatives with respect to q along a curve CQ. Here again a forward/backward sense comes from the underlying field."

Definition (basic derivatives along CQ with respect to q)
Given
CQ = **v**(q), a curve;
f(**v**(q))|CQ, a restricted function;
then if it exists

lim(**f**(**v**(q1+dq)) – **f**(**v**(q1)))/dq as dq ⟶ 0

is called the **basic forward directional derivative of f**|**v**(q) at q1;

lim(**f**(**v**(q1)) – **f**(**v**(q1–dq)))/dq as dq ⟶ 0

is called the **basic backward directional derivative of f**|**v**(q) at q1;
end of definition

Forward derivatives are written as
D[**v**(q),**v**(q+dq)](**f**|CQ;dq)
or D[**v**(q)](**f**|CQ;d_fq)
Backward derivatives are written as
D[**v**(q–dq),**v**(q)](**f**|CQ;dq)
or D[**v**(q)](**f**|CQ;d_bq).

Einstein, questioning: "What makes the derivative basic?"

Breton, answering comprehensively: "We can create functions along CQ with derivatives which are completely arbitrary, even functions without derivatives.

If a function has a *basic* derivatives restricted to CQ at **v**(q1), it is called **differentiable** there. If forward and backward derivatives are equal there, the function is said to have a **continuous** derivative there. If the function has a continuous derivative at each vector of CQ, the function is said to have a continuous restricted derivative **everywhere** along CQ. We can write continuous restricted derivatives as

$$D[\mathbf{v}(q1)](\mathbf{f}(\mathbf{v}(q))|CQ;dq).$$

For such functions

$$\mathbf{f}(\mathbf{v}(q1+dq) \approx \mathbf{f}(\mathbf{v}(q1)) + dq*D[\mathbf{v}(q),\mathbf{v}(q+dq](\mathbf{f}|CQ;dq).$$

Einstein, commandingly: "So next continue to derivatives with respect to **v** along CV!"

Breton: "The way is well marked, but with a surprise."

Definition (basic derivatives along CV with respect to **v**)
Given
CV, a curve with possibly many branches;
v1, a vector in CV;
ut(v1), a curve's direction at **v1**;
f(v)|CV, a restricted function;
for
$\mathbf{dv(v1)} \equiv dq*\mathbf{ut(v1)}$, $dq>0$;
$\mathbf{dv(v1)} = \mathbf{v1} + \mathbf{dv(v1)}$, another vector on a branch in the neighborhood of **v1**;
then if it exists

$$\lim(\mathbf{f}(\mathbf{v1}+\mathbf{dv}(\mathbf{v1})) - \mathbf{f}(\mathbf{v1}))/dq \quad \text{as } dq \rightarrow 0$$

is called a **basic derivative of f(v) at v1 along a branch of CV**.
end of definition

These derivatives are written as

$$D[\mathbf{v1},\mathbf{v1}+\mathbf{dv}](\mathbf{f}(\mathbf{v})|CV;dq)$$

for each branch at **v1**."

Einstein: "Branches! How do we deal with them?"

Breton: "If a function has a basic derivatives restricted to CV at **v1**, it is called **differentiable** in that branch there. If the function has a continuous restricted derivative at each vector of CV along the branch, the function is said to have a continuous restricted derivative **everywhere** along the branch. If the function has a continuous restricted derivative at each vector of CV along all branches at **v1**, the function is said to have a continuous restricted derivative for all branches there. If the function has a continuous restricted derivative at each vector of CV along all branches everywhere in CV, the function is said to have a continuous restricted derivative **everywhere** on CV. When the derivatives for each branch are equal, the derivative is written simply as

$$D[\mathbf{v1}](\mathbf{f}(\mathbf{v})|CV;dq).$$

For such functions

$$\mathbf{f}(\mathbf{v1}+\mathbf{dv}(\mathbf{v1})) \approx \mathbf{f}(\mathbf{v1}) + dq*D[\mathbf{v1},\mathbf{v1}+\mathbf{dv}](\mathbf{f}(\mathbf{v})|CV;dq).$$

Einstein: "All of these derivatives we have been discussing could refer to the same function. They must be related in some way."

Breton: "And so they are."

Theorem: (relationship between derivatives along CQ and CV)

Given

C ≡ CV = CQ = {**v**(q)}, a curve;
v1 = **v**(q1);
ut(**v1**), the direction of C at **v1**;
f(**v**(q)), a compound vector function
with a basic derivative at **v1**;

for

v(q) with a basic derivative at **v1**;
dv(**v1**) ≡ dq***ut**(**v1**) ≡ **v** – **v1**, dq>0;

then

D[**v**(q1),**v**(q1+dq)](**f**(**v**(q))|C;dq)
= abs(D[**v**(q1), q1+dq](**v**(q)|CQ;dq)
*D[**v1**,**v1**+**dv**(**v1**)](**f**(**v**)|CV;dv)

Proof:

Since D[q1, q1+dq](**f**(q)|C;dq) has a basic derivative at q1
f(**v**(q1+dq)) ≈ **f**(**v**(q1)) +dq*D[q1,q1+dq]**f**(q)|C;dq)
Further, for **dv**(**v1**) = **v**(q1 + dq) – **v**(q1)
f(**v**(q1+dq)) = **f**(**v1**+**dv**)
Therefore,
D[q1,q1+dq](**f**(q)|**v**(q);dq)
≈ (**f**(**v**(q1+dq) – **f**(**v**(q1))/dq
≈ ((**f**(**v1** + **dv**(**v1**)) – **f**(**v1**))/dq)*dq/dq)
since dq>0.
Now lim(dq/dq) = abs(D[q1, q1+dq](**v**(q)|C;dq)
Thus
D[q1, q1+dq](**f**(q)|C;dq)
= abs(D[q1, q1+dq](**v**(q)|C;dq)*D[**v1**,**v1**+**dv**(**v1**)](**f**(**v**)|CV;dq)

qed (end of proof)

The theorem gives the conditions for a valid **chain rule** between process derivatives and vectorial derivatives along a curve."

Newton: "What is a chain rule?"

Breton: "We defined a chain rule in tp1.1. If you don't remember let me reiterate:
For a compound variable g(x) which has a continuous, non-zero derivative D[x](g(x)|u;dx)

D[x](f(x)|u;dx) = D[g](f(g)|u;dg)*D[x](g(x)|u;dx)

Please note the conditions. When g and its derivative satisfy the conditions, the above equation is called the **chain rule**.
So the theorem states the conditions for a valid chain rule for curves in **V3**. Chain rules are not always valid."

Einstein, incisively: "Suppose the process is stalled."

Breton: "Then D[q1, q1+dq](**v**(q)|CQ;dq) = **0**. In this case for a non-zero derivative along CQ, the derivative along CV must be unbounded."

Newton: "Better to say that for stalled processes the theorem does not hold."

Einstein, intent on moving the conversation: "Suppose the curve branches."

Breton: "If the curve branches, then at each branch point each of the factors of the theorem may be multi-valued."

Newton, somewhat overwhelmed by the variety of functions: "How about sums of functions?"

Einstein, enforcing the new direction: "Not only sums but for our many vectorial products too?"

Breton: "Good question. Provided we are dealing with functions with basic derivatives, then, vectorial functions, being vectors, conform to our previous results. Let me state these generically, assuming the derivatives apply to the same class of functions.

D[](**f1** + **f2**|u) = D[](**f1**|u) + D[](**f1**|u)
D[](**f1** • **f2**|u) = **f1**•D[](**f2**|u) + **f2**•D[](**f1**|u)
D[](**f1** ∧ **f2**|u) = **f1**∧D[](**f2**|u) + **f2**∧D[](**f1**|u)
D[](**f1** ✱ **f2**|u) = **f1**✱D[](**f2**|u) + **f2**✱D[](**f1**|u).

Einstein: "I'll allow you the sums, but the products you will have to prove."

Breton: "All right. Let

f1 = f11✱**u1** + f12✱**u2** + f13✱**u3**
f2 = f21✱**u1** + f22✱**u2** + f23✱**u3**

and let me start with **f1** ✱ **f2** as the most complicated case which likely will include the others. What is **f1** ✱ **f2**?"

Newton: "That's easy.

f1 ✱ **f2** = f11*f21✱**u1**✱**u1** + f11*f22✱**u1**✱**u2** + f11*f23✱**u1**✱**u3**
+ f12*f21✱**u2**✱**u1** + f12*f22✱**u2**✱**u2** + f12*f23✱**u2**✱**u3**
+ f13*f21✱**u3**✱**u1** + f13*f22✱**u3**✱**u2** + f13*f23✱**u3**✱**u3**.

Breton: "Now the derivative of any element of the transformation, say fi2*f2j✱**ui**✱**uj** is

D[](fi2*f2j|u)✱**ui**✱**uj** = fi2*D[](f2j|u)✱**ui**✱**uj**
+ f2j*D[](fi2|u)✱**ui**✱**uj**.

Einstein: "Only if those derivatives exist."

Breton: "Yes, of course. They do exist for differentiable functions. So the derivative of an outer product can be split in two. Now let us look at the first part of the split

$$
\begin{aligned}
& f11*D[\,](f21|u)*\mathbf{u1}*\mathbf{u1} \\
& \quad + f11*D[\,](f22|u)*\mathbf{u1}*\mathbf{u2} \\
& \quad + f11*D[\,](f23|u)*\mathbf{u1}*\mathbf{u3} \\
& \quad + f12*D[\,](f21|u)*\mathbf{u2}*\mathbf{u1} \\
& \quad + f12*D[\,](f22|u)*\mathbf{u2}*\mathbf{u2} \\
& \quad + f12*D[\,](f23|u)*\mathbf{u2}*\mathbf{u3} \\
& \quad + f13*D[\,](f21|u)*\mathbf{u3}*\mathbf{u1} \\
& \quad + f13*D[\,](f22|u)*\mathbf{u3}*\mathbf{u2} \\
& \quad + f13*D[\,](f23|u)*\mathbf{u3}*\mathbf{u3} \\
& \qquad = f11*\mathbf{u1}*D[\,](f21|u)*\mathbf{u1} \\
& \qquad\qquad + f11*\mathbf{u1}*D[\,](f22|u)*\mathbf{u2} \\
& \qquad\qquad + f11*\mathbf{u1}*D[\,](f23|u)*\mathbf{u3} \\
& \qquad + f12*\mathbf{u2}*D[\,](f21|u)*\mathbf{u1} \\
& \qquad\qquad + f12*\mathbf{u2}*D[\,](f22|u)*\mathbf{u2} \\
& \qquad\qquad + f12*\mathbf{u2}*D[\,](f23|u)*\mathbf{u3} \\
& \qquad + f13*\mathbf{u3}*D[\,](f21|u)*\mathbf{u1} \\
& \qquad\qquad + f13*\mathbf{u3}*D[\,](f22|u)*\mathbf{u2} \\
& \qquad\qquad + f13*\mathbf{u3}*D[\,](f23|u)*\mathbf{u3}.
\end{aligned}
$$

Newton: "That is just the expansion of $\mathbf{f1}*D[\,](\mathbf{f2}|u)$."

Breton: "Similarly for the second element $\mathbf{f2}*D[\,](\mathbf{f1}|u)$. So does this prove

$$D[\,](\mathbf{f1} * \mathbf{f2}|u) = \mathbf{f1}*D[\,](\mathbf{f2}|u) + \mathbf{f2}*D[\,](\mathbf{f1}|u)?"$$

Both Newton and Einstein nod in agreement, Newton with somewhat more enthusiasm.

Breton: "Since inner products and cross products are related to outer products as traces and curl matrix operator of outer products, dot and cross products are proven similarly."

Newton: "And if one of the functions is constant, that is with a derivative of zero, then for a constant vector **c**

$$
\begin{aligned}
& D[\,](\mathbf{c} \bullet \mathbf{f}|u) = \mathbf{c}\bullet D[\,](\mathbf{f}|u) \\
& D[\,](\mathbf{c} \wedge \mathbf{f}|u) = \mathbf{c}\wedge D[\,](\mathbf{f}|u) \\
& D[\,](\mathbf{c} * \mathbf{f}|u) = \mathbf{c}*[\,](\mathbf{f}|u).
\end{aligned}
$$

Einstein: "Can you give us an example?"

Breton, accepting the request: "Since

$$(\mathbf{f}|u)\bullet(\mathbf{f}|u) = abs(\mathbf{f}|u)*abs(\mathbf{f}|u),$$

$$\mathbf{f}|u)\bullet D[\,](\mathbf{f}|u) = abs(\mathbf{f}|u)*D[\,](abs(\mathbf{f}|u))$$

even when $D[\,](abs(\mathbf{f}|u))$ does *not* equal $abs(D[\,](\mathbf{f}|u))$.

Furthermore, if abs(**f**|u) is constant (even though **f** varies)

D[](**f** • **f**|u) = 2•**f**|u)•D[](**f**|u)

= 0

so that in this case **f** and D[](**f**|u) are orthogonal."

Newton: "What kind of example is that?"

Breton: "Think of a function restricted to the surface of a sphere, and thus constant in magnitude. The function may change in direction, but its magnitude will remain constant. Now we can see that for such functions with basic derivatives, the function itself and its derivative are orthogonal."

Einstein, intent on leading by moving forward: "Good. How about the vectorial derivatives like divergences?"

Directional Divergences, Curls, and Gradients

Breton: "We saw that the directional reciprocal vector

qd(**v**) ≡ **u**(**v**)/v

played a key role in the solution of vectorial equations containing inner products. It becomes a key factor in the definition of directional gradients."

Einstein: "Don't keep us in suspense. Give us a definition!"

Breton: "Sure, here it is."

Definition (basic directional divergences, curls, and gradients)

Given

uv ≡ **u**(**v**–v**1**), a direction;
dv ≡ **v**–**v1**;
f(**v**) | **uv**, a restricted function;

then

the basic forward directional divergence, curl, and gradient in the direction **uv** are defined at **v1** as

Divergence:
D[**v1**,**v1**+**dv**] • (**f**(**v**) | **uv**;**dv**) ≡ lim(**f**(**v1**+**dv**)) – **f**(**v1**)) • **qd**(**v**)

Curl:
D[**v1**,**v1**+**dv**] ∧ (**f**(**v**) | **uv**;**dv**) ≡ lim(**f**(**v1**+**dv**)) – **f**(**v1**)) ∧ **qd**(**v**)

Gradient:
D[**v1**,**v1**+**dv**] ⁕ (**f**(**v**) | **uv**;**dv**) ≡ lim(**f**(**v1**+**dv**)) – **f**(**v1**)) ⁕ **qd**(**v**)

the basic backward directional divergence, curl, and gradient in the direction **uv** are defined at **v1** as

Divergence:
D[**v1**–**dv**,**v1**] • (**f**(**v**) | **uv**;**dv**) ≡ lim(**f**(**v1**) – **f** (**v1**–**dv**)) • **qd**(**v**)

Curl:
D[**v1**–**dv**,**v1**] ∧ (**f**(**v**) | **uv**;**dv**) ≡ lim(**f**(**v1**) – **f** (**v1**–**dv**)) ∧ **qd**(**v**)

Gradient:
D[**v1**–**dv**,**v1**] ⁕ (**f**(**v**) | **uv**;**dv**) ≡ lim(**f**(**v1**) – **f** (**v1**–**dv**)) ⁕ **qd**(**v**)

as **dv** ⟶ **0**

end of definition

Einstein: "Then we can also define these derivatives along CV."

Breton: "Yes, the path here is clear enough."

Definition (basic divergences, curls, and gradients along CV)

Given

CV, a curve in **V3**;

v1, a vector in CV;

ut(v1), the direction of CV at **v1**;

f(v)|CV, a restricted function;

for

dv(v1) ≡ dv ✱ **ut(v1)**, dv>0;

then

the basic forward directional divergence, curl, and gradient along CV are defined at **v1** as

Divergence:

D[**v1**,**v1**+**dv**(**v1**)] • (**f**(**v**)|CV;**dv**)

≡ lim(**f**(**v1**+**dv**(**v1**)) – **f**(**v1**))) • **qd**(**dv**(**v1**))

Curl:

D[**v1**,**v1**+**dv**(**v1**)]∧(**f**(**v**)|CV;**dv**)

≡ lim(**f**(**v1**+**dv**(**v1**)) – **f**(**v1**)))∧**qd**(**dv**(**v1**))

Gradient:

D[**v1**,**v1**+**dv**(**v1**)]✱(**f**(**v**)|CV;**dv**)

≡ lim(**f**(**v1**+**dv**(**v1**)) – **f**(**v1**)))✱**qd**(**dv**(**v1**))

as dv ⟶ **0**

end of definition

Newton: "Then basic backward derivatives along CV may be defined analogously."

Breton, somewhat dismissively while pressing onward: "Yes. If a function has a basic restricted divergence, curl, and gradient it is called **differentiable** according to the restriction. If forward and backward divergences, curls, and gradients are equal at some reference vector, the function is said to have a **continuous** restricted derivative according to the restriction there. A function may have a continuous restricted divergence, curl and gradient **everywhere** in the restriction."

Newton, undismissed: "What symbolism can we use when the forward and backward derivatives are equal?"

Breton: "Then the derivatives are written more simply as

D[**v1**,**v1**+**dv**] • (**f**(**v**)|C;**dv**)

D[**v1**,**v1**+**dv**]∧(**f**(**v**)|C;**dv**)

D[**v1**,**v1**+**dv**]✱(**f**(**v**)|C;**dv**).

Note that **D[v1,v1+dv]•(f(v)**|C;**dv**) is a transformation, specifically and outer product whose rank is thus 1."

Einstein: "When we first symbolized derivatives in tp1.1 we talked about derivatives as functions of their references which may also vary. Why not with these derivatives also?"

Breton: "Right on. The *everywhere* condition may be expressed as a variable. Then the derivatives become functions of **v** and are written, for example, as

D[v,v+dv]•(f(v)|C;**dv**).

Newton, taking the lead: "Just now we talked about sums of these derivatives. It strikes me that triple products can make a mark here."

Breton: "That's an interesting topic. What can we learn from forming triple products with derivatives? Consider, for a first instance,

((**f(v(**q1+dq)) – **f(v(**q1)))/dq) • (**d(v(**q1))∧**qd(d(v(**q1))))

as a scalar triple product."

Newton, taking up the suggestion: "As limiting forms

((**f(v(**q1+dq)) – **f(v(**q1)))/dq)⟶D[**v**(q1),**v**(q1+dq)](**f**|CQ;dq)

(**d(v(**q1))∧**qd(d(v(**q1))))⟶**D**[**v**(q1),**v(**q1+dq)]∧(**v**(q)|CQ;dq).

Einstein, dismissively: "I note that (**d(v(**q))∧**qd(d(v(**q)))) = **0**."

Breton, proceeding undisturbed: "Yes it does. Do you note also

((**f(v(**q1+dq)) – **f(v(**q1)))/dq) • (**d(v(**q1))∧**qd(d(v(**q1))))
= –((**f(v(**q1+dq))–**f(v(**q1)))/dq) • (**qd(d(v(**q1))∧**d(v(**q1))))
= –((**f(v(**q1+dq))–**f(v(**q1)))/dq)∧(**qd(d(v(**q1))) • **d(v(**q1)))
= –((**f(v(**q1+dq))–**f(v(**q1)))∧**qd(d(v(**q1)))) • (**d(v(**q1))/dq).

Newton, concluding: "As limiting forms

(((**f(v(**q1+dq)) – **f(v(**q1)))∧(**qd(d(v(**q))
⟶**D**[**v**(q1),**v**(q1+dq)]∧(**f(v)**|CQ;**dv**)

d(v(q1))/dq)⟶D[q1,q1+dq](**v**((q);dq).

Einstein, accentuating: "Also equal to **0**."

Breton, concluding: "Thus we discover

D[**v**(q1),**v**(q1+dq)]∧(**f(v)**|CQ;**dv**)
and D[q1,q1+dq](**v**((q);dq)
are orthogonal vectors."

Newton, in astonishment: "Remarkable."

Einstein, recovering: "Can we discover something similar with the vector triple product?"

Breton: "Consider again.
((f(v(q1+dq) – f(v(q1)))/dq)∧(d(v(q1))∧qd(d(v(q1))))
= ((f(v(q1+dq) – f(v(q1)))•qd(d(v(q1))))∗(d(v(q1))/dq)
– (((f(v(q1+dq) – f(v(q1)))/dq)•d(v(q1)))∗qd(d(v(q1))).
Then as limiting forms
((f(v(q1+dq) – f(v(q1)))•qd(d(v(q))))
→D[v(q1),v(q1+dq)]•(f|CQ;dv)
(d(v(q1))/dq)→D[q1,q1+dq](v((q);dq)
((f(v(q1+dq) – f(v(q1)))/dq)→D[v(q1),v(q1+dq)](f|CQ;dq)
qd(d(v(q)))∗d(v(q))→ut(q1)∗ut(q1).
So taking up your observation Einstein
D[v(q1),v(q1+dq)]•(f(v)|CQ;dv)∗D[q1,q1+dq](v(q);dq)
– D[v(q1),v(q1+dq)](f(v)|CQ;dq)•ut(q1)∗ut(q1)
= 0.
Consequently
D[v(q1),v(q1+dq)]•(f(v)|CQ;dv)∗D[q1,q1+dq](v(q);dq)
= D[v(q1),v(q1+dq)](f(v)|CQ;dq)•ut(q1)∗ut(q1).

Einstein, now amazed himself: "Truly remarkable. These products have a specific direction."

Newton: "You've given us quite a mouthful."

Breton: "There is still more. Consider again
((f(v(q1+dq)) – f(v(q1)))/dq)∗(d(v(q1))∧qd(d(v(q1))))
= –((f(v(q1+dq))–f(v(q1)))∗qd(d(v(q1))))•C(d(v(q1)/dq).

Newton: "Where do you get this from?"

Breton: "Check with the table of vectorial identities you assembled for tp1.2. In this form
D[v(q1),v(q1+dq)](f(v)|CQ;dq)∗D[q1,q1+dq]∧(v((q)|CQ;dv)
–D[v(q1),v(q1+dq)]∗(f(v)|CQ;dv)•C(D[q1,q1+dq](v(q);dq))
= 0.

Newton: "So we could possibly prove more identities using the identities from tp1.2."

Breton: "Possibly. Please note that the functions need not be ones with basic derivatives. Only for those with basic derivatives can we write
f(v(q1+dq) ≈ f(v(q1)) + dq∗D[v(q1),v(q1+dq](f|CQ;dq)
f(v(q1+dq) ≈ f(v(q1)) + dv•D[v(q1),v(q1+dq]∗(f|CQ;dv)
f(v(q1+dq) ≈ f(v(q1)) + dv∗D[v(q1),v(q1+dq]•(f|CQ;dv)
f(v1+dv∗uv) ≈ f(v1) + dv∗D[v1,v1+dq∗uv](f(v)|CV;dv))
f(v1+dv) ≈ f(v1) + dv•D[v1,v1+dv]∗(f(v)|CV;dv))
f(v1+dv) ≈ f(v1) + dv∗D[v1,v1+dv]•(f(v)|CV;dv)).

Einstein, again leading by questioning: "How about derivatives based on other limits?"

First quadrant derivatives

Breton: "Let's continue on then to first quadrant derivatives. For these derivatives, the reciprocal vector $\mathbf{q}(\mathbf{v})$ becomes a central factor."

Newton: "Remind us about $\mathbf{q}(\mathbf{v})$."

Breton, complying: " For any

$$\begin{aligned} \mathbf{v} &= v * \mathbf{uv} \\ &= v1 * \mathbf{u1} + v2 * \mathbf{u2} + v3 * \mathbf{u3} \\ &= v * (c1 * \mathbf{u1} + c2 * \mathbf{u2} + c3 * \mathbf{u3}) \end{aligned}$$

$$\begin{aligned} \mathbf{q}(\mathbf{v}) &\equiv (\mathbf{u1}/v1 + \mathbf{u2}/v2 + \mathbf{u3}/v3) \\ &= (\mathbf{u1}/c1 + \mathbf{u2}/c2 + \mathbf{u3}/c3)/v \\ &= \mathbf{q}(\mathbf{uv})/v. \end{aligned}$$

Note the differences between $\mathbf{q}(\mathbf{v}) = \mathbf{q}(\mathbf{uv})/v$ and $\mathbf{qd}(\mathbf{v}) = \mathbf{uv}/v$.

Einstein: "How about some definitions?"

Breton: "Certainly.

> **Definition** (basic positive quadrant divergences, curls, and gradients)
>
> Given
>
> quotient numbers $dvi > 0$, $i=1,2,3$;
>
> $\mathbf{dv} = dv1 \bullet \mathbf{u1} + dv2 \bullet \mathbf{u2} + dv3 \bullet \mathbf{u3}$;
>
> $\mathbf{v1}$, a vector in $\mathbf{V3}$;
>
> $\mathbf{f}(\mathbf{v}) = \mathbf{f}(v1 * \mathbf{u1} + v2 * \mathbf{u2} + v3 * \mathbf{u3})$;
>
> for
>
> tr, the trace matrix operator;
>
> c, the curl matrix operator;
>
> G, the diagonal vector operator;
>
> then
>
> $$\begin{aligned} \lim\, &(\mathbf{f}(\mathbf{v1} + dv1 \bullet \mathbf{u1} - \mathbf{f}(\mathbf{v1})) \bullet \mathbf{u1}/dv1 \\ &+ (\mathbf{f}(\mathbf{v1} + dv2 \bullet \mathbf{u2} - \mathbf{f}(\mathbf{v1})) \bullet \mathbf{u2}/dv2 \\ &+ (\mathbf{f}(\mathbf{v1} + dv3 \bullet \mathbf{u3} - \mathbf{f}(\mathbf{v1})) \bullet \mathbf{u3}/dv3 \\ = \lim\, &(((\mathbf{f}(\mathbf{v1} + dv1 \bullet \mathbf{u1} - \mathbf{f}(\mathbf{v1})) \bullet \mathbf{u1} \bullet \mathbf{u1} \\ &+ (\mathbf{f}(\mathbf{v1} + dv2 \bullet \mathbf{u2} - \mathbf{f}(\mathbf{v1})) \bullet \mathbf{u2} \bullet \mathbf{u2} \\ &+ (\mathbf{f}(\mathbf{v1} + dv3 \bullet \mathbf{u3} - \mathbf{f}(\mathbf{v1})) \bullet \mathbf{u3} \bullet \mathbf{u3}) \bullet \mathbf{q}(\mathbf{dv}) \\ = \lim\, &(tr[(\mathbf{f}(\mathbf{v1} + dv1 \bullet \mathbf{u1} - \mathbf{f}(\mathbf{v1})) \bullet \mathbf{u1} \bullet \mathbf{u1} \\ &+ (\mathbf{f}(\mathbf{v1} + dv2 \bullet \mathbf{u2} - \mathbf{f}(\mathbf{v1})) \bullet \mathbf{u2} \bullet \mathbf{u2} \\ &+ (\mathbf{f}(\mathbf{v1} + dv3 \bullet \mathbf{u3} - \mathbf{f}(\mathbf{v1})) \bullet \mathbf{u3} \bullet \mathbf{u3})] \bullet [\mathbf{G}(\mathbf{q}(\mathbf{dv}))] \\ &\qquad \text{as } \mathbf{dv}+ \longrightarrow \mathbf{0} \end{aligned}$$
>
> is called the **positive quadrant divergence** of $\mathbf{f}(\mathbf{v})$ at $\mathbf{v1}$.

then also

$$\begin{aligned}
&\lim (\mathbf{f}(\mathbf{v1} + dv1\bullet\mathbf{u1} - \mathbf{f}(\mathbf{v1}))\wedge\mathbf{u1}/dv1 \\
&\qquad + (\mathbf{f}(\mathbf{v1} + dv2\bullet\mathbf{u2} - \mathbf{f}(\mathbf{v1}))\wedge\mathbf{u2}/dv2 \\
&\qquad + (\mathbf{f}(\mathbf{v1} + dv3\bullet\mathbf{u3} - \mathbf{f}(\mathbf{v1}))\wedge\mathbf{u3}/dv3 \\
&\quad = \lim ((\mathbf{f}(\mathbf{v1} + dv1\bullet\mathbf{u1} - \mathbf{f}(\mathbf{v1}))\cdot(\mathbf{u3}\bullet\mathbf{u2} - \mathbf{u2}\bullet\mathbf{u3})/dv1 \\
&\qquad + ((\mathbf{f}(\mathbf{v1} + dv2\bullet\mathbf{u2} - \mathbf{f}(\mathbf{v1}))\cdot(\mathbf{u1}\bullet\mathbf{u3} - \mathbf{u3}\bullet\mathbf{u1})/dv2 \\
&\qquad + ((\mathbf{f}(\mathbf{v1} + dv3\bullet\mathbf{u3} - \mathbf{f}(\mathbf{v1}))\cdot(\mathbf{u2}\bullet\mathbf{u1} - \mathbf{u1}\bullet\mathbf{u2})/dv3 \\
&\quad = \lim (\mathbf{f}(\mathbf{v1} + dv3\bullet\mathbf{u3} - \mathbf{f}(\mathbf{v1}))/dv3\cdot\mathbf{u2} \\
&\qquad\qquad - (\mathbf{f}(\mathbf{v1} + dv2\bullet\mathbf{u2} - \mathbf{f}(\mathbf{v1}))/dv2\cdot\mathbf{u3})\bullet\mathbf{u1} \\
&\qquad + (\mathbf{f}(\mathbf{v1} + dv1\bullet\mathbf{u1} - \mathbf{f}(\mathbf{v1}))/dv1\cdot\mathbf{u3} \\
&\qquad\qquad - (\mathbf{f}(\mathbf{v1} + dv3\bullet\mathbf{u3} - \mathbf{f}(\mathbf{v1}))/dv3\cdot\mathbf{u1})\bullet\mathbf{u2} \\
&\qquad + (\mathbf{f}(\mathbf{v1} + dv2\bullet\mathbf{u2} - \mathbf{f}(\mathbf{v1}))/dv2\cdot\mathbf{u1} \\
&\qquad\qquad - (\mathbf{f}(\mathbf{v1} + dv1\bullet\mathbf{u1} - \mathbf{f}(\mathbf{v1}))/dv1\cdot\mathbf{u2})\bullet\mathbf{u3} \\
&\quad = \lim \mathbf{c}[((\mathbf{f}(\mathbf{v1} + dv1\bullet\mathbf{u1} - \mathbf{f}(\mathbf{v1}))\bullet\mathbf{u1} \\
&\qquad\qquad + (\mathbf{f}(\mathbf{v1} + dv2\bullet\mathbf{u2} - \mathbf{f}(\mathbf{v1}))\bullet\mathbf{u2} \\
&\qquad\qquad + (\mathbf{f}(\mathbf{v1} + dv3\bullet\mathbf{u3} - \mathbf{f}(\mathbf{v1}))\bullet\mathbf{u3}]\cdot[\mathbf{G}(\mathbf{q}(\mathbf{dv}))] \\
&\qquad\qquad\qquad \text{as } \mathbf{dv+} \longrightarrow \mathbf{0}
\end{aligned}$$

is called the **positive quadrant curl** of **f(v)** at **v1**.

then also for a scalar function f

$$\begin{aligned}
&\lim (f(\mathbf{v1} + dv1\bullet\mathbf{u1} - f(\mathbf{v1}))\bullet\mathbf{u1}/dv1 \\
&\qquad + (f(\mathbf{v1} + dv2\bullet\mathbf{u2} - f(\mathbf{v1}))\bullet\mathbf{u2}/dv2 \\
&\qquad + (f(\mathbf{v1} + dv3\bullet\mathbf{u3} - f(\mathbf{v1}))\bullet\mathbf{u3}/dv3 \\
&\quad = \lim [(f(\mathbf{v1} + dv1\bullet\mathbf{u1} - f(\mathbf{v1}))\bullet\mathbf{u1} \\
&\qquad\qquad +(f(\mathbf{v1} + dv2\bullet\mathbf{u2} - f(\mathbf{v1}))\bullet\mathbf{u2} \\
&\qquad\qquad +(f(\mathbf{v1} + dv3\bullet\mathbf{u31} - f(\mathbf{v1}))\bullet\mathbf{u3}]\cdot[\mathbf{G}(\mathbf{q}(\mathbf{dv}))] \\
&\qquad\qquad\qquad \text{as } \mathbf{dv+} \longrightarrow \mathbf{0}
\end{aligned}$$

is called the **positive quadrant gradient** of **f(v)** at **v1**.

end of definition

The positive quadrant divergence of a vector function is symbolized as

$$D[\mathbf{v1}]\cdot(\mathbf{f}(\mathbf{v});\mathbf{dv})$$

The positive quadrant curl of a vector function is symbolized as

$$\mathbf{D}[\mathbf{v1}]\wedge(\mathbf{f}(\mathbf{v});\mathbf{dv})$$

The positive quadrant gradient of a scalar function is symbolized as

$$\mathbf{D}[\mathbf{v1}]\bullet(f(\mathbf{v});\mathbf{dv}).$$

Newton: "These derivatives look something like directional derivatives."

Breton: "More than looks. They can be expressed in terms of directional derivatives as:

$$\begin{aligned}
D[\mathbf{v1}]\cdot(\mathbf{f}(\mathbf{v});\mathbf{dv}) = {} & D[\mathbf{v1}, dv1\bullet\mathbf{u1}](\mathbf{f}(\mathbf{v});dv1)\cdot\mathbf{u1} \\
& +D[\mathbf{v1}, dv2\bullet\mathbf{u2}](\mathbf{f}(\mathbf{v});dv1)\cdot\mathbf{u2} \\
& +D[\mathbf{v1}, dv3\bullet\mathbf{u3}](\mathbf{f}(\mathbf{v});dv1)\cdot\mathbf{u3}
\end{aligned}$$

D[**v1**]∧(**f**(**v**);**dv**) = D[**v1**, dv1•**u1**](**f**(**v**);dv1)∧**u1**
+D[**v1**, dv2•**u2**](**f**(**v**);dv1)∧**u2**
+D[**v1**, dv3•**u3**](**f**(**v**);dv1)∧**u3**
D[**v1**]•(f(**v**);**dv**) = D[**v1**, dv1•**u1**](f(**v**);dv1)•**u1**
+D[**v1**, dv2•**u2**](f(**v**);dv1)•**u2**
+D[**v1**, dv3•**u3**](f(**v**);dv1)•**u3.**

These derivatives differ from functional limits of the positive quadrant inasmuch as the latter have the same limit in the three orthogonal directions."

Newton, objecting: "But these may not be continuous in the negative direction."

Breton: "Correct."

Einstein, also objecting: "These definitions depend on the given orientation. In a different orientation they may be different."

Breton: "Correct again. The limit may not even exist in a different orientation. The form of the a gradient, in common with all local derivatives, depends on the orientation of the origin, but not necessarily on the value of the function. As the attribute of order weakens, a modified definition of continuity will be needed."

Einstein: "Show us."

Breton: "First let me continue with other positive quadrant ideas. If a function has a basic positive quadrant divergence, curl and gradient at **v1**, it is call **differentiable** there.

Do you think a positive quadrant gradient can be defined for **vector** functions?"

Newton: "Why not?"

Breton: "The definitions of divergences, curls, and gradients all involve vectors as either inputs or outputs or both. The gradient of a vector function, however, would yield a transformation."

Newton: "Still we can try. How do we start?"

Breton: "Let's start with a function **f**(**v**) = **c**•f(**v**) where **c** is a constant vector and f(**v**) is a scalar function. The formal application of the definition of a gradient yields the result

D[**v1**]•(**c**•(f(**v**);**dv**)) = **c**•**D**[**v1**]•(f(**v**);**dv** + f(**v**)•**D**[**v1**]•(**c**;**dv**)
= **c**•**D**[**v1**]•(f(**v**);**dv**

since for a constant vector **D**[**v1**]•(**c**•;**dv**)) would be [**0**}.

Newton, observing: "So the result is an outer product."

Einstein, observing further: "Which does not equal
D[**v1**]•(f(**v**);**dv**)•**c**."

Newton, skeptically: "Where does this lead us?"

Breton: "Any non-restricted vector function can be expressed as

$$\begin{aligned}\mathbf{f(v)} &= \mathbf{f(v)\cdot u1\bullet u1 + f(v)\cdot u2\bullet u2 + f(v)\cdot u3\bullet u3} \\ &\equiv \mathrm{f1}(\mathbf{v})\bullet\mathbf{u1} + \mathrm{f2}(\mathbf{v})\bullet\mathbf{u2} + \mathrm{f3}(\mathbf{v})\bullet\mathbf{u3}.\end{aligned}$$

Newton, enlightened: "We know how to compute the gradient of each addend. The gradient of the entire function would then become

$$\begin{aligned}\mathbf{D[v1]\bullet(f(v);dv)} = \mathbf{u1\bullet D[v1]\bullet}(\mathrm{f1}(\mathbf{v});\mathbf{dv}) \\ + \mathbf{u2\bullet D[v1]\bullet}(\mathrm{f2}(\mathbf{v});\mathbf{dv}) \\ + \mathbf{u3\bullet D[v1]\bullet}(\mathrm{f3}(\mathbf{v});\mathbf{dv}).\end{aligned}$$

Breton: "Take note that $\mathrm{f1}(\mathbf{v}) = \mathbf{f(v)\cdot u1}$."

Newton: "Then

$$\begin{aligned}\mathbf{D[v1]\bullet(f(v);dv)} = \mathbf{u1\bullet D[v1]\bullet(f(v)\cdot u1;dv)} \\ + \mathbf{u2\bullet D[v1]\bullet(f(v)\cdot u2;dv)} \\ + \mathbf{u3\bullet D[v1]\bullet(f(v)\cdot u3;dv)}.\end{aligned}$$

Einstein, having followed he development closely: "Which we can then rewrite as

$$\begin{aligned}\mathbf{D[v1]\bullet(f(v);dv)} = \mathbf{u1\bullet D[v1]\bullet(u1\cdot f(v);dv)} \\ + \mathbf{u2\bullet D[v1]\bullet(u2\cdot f(v);dv)} \\ + \mathbf{u3\bullet D[v1]\bullet(u3\cdot f(v);dv)}.\end{aligned}$$

Newton: "Which can be further rewritten as

$$\begin{aligned}\mathbf{D[v1]\bullet(f(v);dv)} = \mathbf{u1\bullet u1\cdot D[v1]\bullet(f(v);dv)} \\ + \mathbf{u2\bullet u2\cdot D[v1]\bullet(f(v);dv)} \\ + \mathbf{u3\bullet u3\cdot D[v1]\bullet(f(v);dv)}.\end{aligned}$$

Breton: "Which is a tautology since

$$\mathbf{u1\bullet u1+u2\bullet u2+u3\bullet u3 = I},$$

the identity matrix.

Newton: "Tautology it may be, but we have the makings of a definition here. I propose

$$\begin{aligned}&\lim\ ((\mathbf{f(v1}+\mathrm{dv1}\bullet\mathbf{u1)} - \mathbf{f(v1)})\bullet\mathbf{u1} \\ &\quad + (\mathbf{f(v1}+\mathrm{dv2}\bullet\mathbf{u2)} - \mathbf{f(v1)})\bullet\mathbf{u2} \\ &\quad + (\mathbf{f(v1}+\mathrm{dv3}\bullet\mathbf{u3)} - \mathbf{f(v1)})\bullet\mathbf{u3})\cdot[\mathbf{G(q(dv))}] \quad \text{as } \mathbf{dv+} \longrightarrow \mathbf{0}\end{aligned}$$

as the definition of $\mathbf{D[v1]\bullet(f(v);dv)}$."

Breton: "Your definition accords beautifully with our previous definitions."

Einstein, inquisitively: "How many derivatives are involved in the this definition?"

Newton: "It looks like only three."

Breton: "Looks can be deceiving. The definition shows the gradient of a vector can be seen as the multiplication of two 3x3 matrices. Look at the first row of the first matrix. It is an outer product

(f(v1+dv1**•u1)** – **f(v1))•u1**

which, in a matrix setting, is made up of three components in the first row. So also for the other two rows. We see then that the first matrix is composed of nine elements. The second matrix is a diagonal matrix, so the matrix multiplication still yields *nine* elements, each of which is a derivative."

Einstein: "So we have arrived at the definition for the positive quadrant gradient of a vector function. What can you say about it?"

Breton: "A few things more. First what is **D[v1]•(v;dv)**?"

Newton: "I find it is just **u1•u1+u2•u2+u3•u3 = I** the identity matrix."

Breton: "What do you make of the trace of the gradient?"

Newton: "I find

tr[**D[v1]•(f(v);dv)**] = **D[v1]•(f(v);dv)**

which is the first quadrant divergence of the function."

Breton: "What do you make of the curl matrix operator of the gradient?"

Newton: "Remind me about the curl matrix operator."

Breton: "Remember in tp1.2 the curl matrix operator of a matrix **A** is defined as

c[A] ≡ (a23–a32)∗**u1** + (a31–a13)∗**u2** + (a12–a21)∗**u3**."

With that Newton set to work and in a few minutes reported: "I find

c[D[v1]•(f(v);dv)] = **D[v1]∧(f(v);dv)**

which is the first quadrant curl of the function."

Breton: "How about the curl matrix operator of the transpose of the gradient?"

Newton: "I find

c[T[D[v1]•(f(v);dv)]] = –**D[v1]∧(f(v);dv)**.

Breton: "There are others too. We can use the vectorial identities we have established in tp1.2 to form more identities."

Einstein: "Enough. I see the positive quadrant gradient is a unifying idea which embraces divergences, curls and scalar gradients."

Breton: "Exactly. We can proceed more efficiently up our mountain of discovery by concentrating on vectorial gradients, knowing that our results also apply analogously to divergences and curls."

Newton: "Put down some specific results."

Breton: "In addition to **D[v1]•(v;dv) = I**
we can write

D[v1]•(v;dv) = 3
D[v1]∧(v;dv) = 0.

Newton: "How about other specific functions?"

Breton: "Here are a few which I leave you to prove for yourself. Otherwise, trust me. The symbols **v1**, **uvv1**, and **A** refer to a constant vector, the direction of **(v–v1)**, and a fixed transformation respectively."

Function	D•	**D∧**	**D***
uvv1 = **(v–v1)** /abs(**v–v1**)	2/abs(**v–v1**)	**0**	**(I–uvv1*uvv1)** /abs(**v–v1**)
v–v1	3	**0**	**I**
(v–v1)•(v–v1)			2***(v–v1)**
abs(**v–v1**)			**uvv1**
(abs(**v–v1**))^n			**uvv1** *(abs(**v–v1**))$^{(n-1)}$
1/abs(**v–v1**)			**–uvv1** /**(v–v1)•(v–v1)**
uvv1 /**(v–v1)•(v–v1)**	0	**0**	**(I–3*uvv1)*uvv1)**/ (abs(**v–v1**))^3
f(**v**)***(v–v1)**	3*f(**v**) + **(v–v1)•D***f	**(v–v1)∧D***f	f(**v**)***I** +**(v–v1)*D***f
(v–v1)•A	tr[**A**]	c[**A**]	**T[A]**
v1•A•v			**v1•A**
v•A•v1			**v1•T[A]**
(v–v1)•A•(v–v1)			**(v–v1)•[A+T[A]**

Some specific positive quadrant results.

Einstein: "Shouldn't these results be restricted to basic derivatives?"

Breton: "Thank you Einstein. Yes, of course. Functions with the same functional values may have different positive quadrant gradients, or even none at all. Furthermore, even functions with basic directional derivatives in the orthogonal directions of the first quadrant may not relate to gradients in other quadrants."

Newton: "How can we deal with those functions?"

Breton: "We need an expanded definition of differentiability."

Definition (continuous vectorial differentiability)
Given
u1, **u2**, **u3**, the orthogonal orientation of the origin;
dv1, dv2, dv3, quotient numbers;
dv ≡ dv1•**u1** +dv2•**u2** +dv3•**u3**;
udv, the direction of **dv**;
f(v), a function over **V3**;
if for every **udv**

$$\mathbf{f}(\mathbf{v1}+\mathbf{dv}) - \mathbf{f}(\mathbf{v1}) = \mathbf{f}(\mathbf{v1}+dv1\bullet\mathbf{u1}) - \mathbf{f}(\mathbf{v1}) + \mathbf{f}(\mathbf{v1}+dv2\bullet\mathbf{u2}) - \mathbf{f}(\mathbf{v1}) + \mathbf{f}(\mathbf{v1}+dv3\bullet\mathbf{u3}) - \mathbf{f}(\mathbf{v1}) \quad \text{as } \mathbf{dv}\to 0$$

then
f(v) is **continuously differentiable vectorially** at **v1**.
end of definition

Newton, inquiring: "Suppose **f** is vectorially differentiable for all vectors in a subset of **V3**?"

Breton, taking Newton's lead: "Call the subset S. Then **f** is said to be vectorially differentiable on S. And if the function is locally differentiable only for positive dvi, the function is said to locally differentiable at **v1** in the positive quadrant."

Einstein, taking up the narrative: "Wherever a function is vectorially differentiable at **v1** in the positive quadrant, the existence of limits implies it has a positive quadrant gradient there. The definition of a positive gradient may be applied to functions which are continuously differentiable vectorially as:

$$\mathbf{D}[\mathbf{v1}]\bullet(\mathbf{f}(\mathbf{v});\mathbf{dv}). \equiv \lim\, [(\mathbf{f}(\mathbf{v1} + dv1\bullet\mathbf{u1} - \mathbf{f}(\mathbf{v1}))\bullet\mathbf{u1} + (\mathbf{f}(\mathbf{v1} + dv2\bullet\mathbf{u2} - \mathbf{f}(\mathbf{v1}))\bullet\mathbf{u2} + (\mathbf{f}(\mathbf{v1} + dv3\bullet\mathbf{u31} - \mathbf{f}(\mathbf{v1}))\bullet\mathbf{u3}]\bullet[\mathbf{G}(\mathbf{q}(\mathbf{dv}))] \quad \text{as } \mathbf{dv}+ \longrightarrow \mathbf{0}$$

Surely this applies only to functions with a basic gradient."

Breton: "Yes, of course. A function with the same functional values may have different positive quadrant gradients even if is vectorially differentiable. For functions with a basic gradient

$$\begin{aligned}\mathbf{f}(\mathbf{v1}+\mathbf{dv}) - \mathbf{f}(\mathbf{v1}) \approx\ & \mathbf{dv}\bullet(\mathbf{u1}\bullet(\mathbf{f}(\mathbf{v1} + \mathrm{dv1}\bullet\mathbf{u1} - \mathbf{f}(\mathbf{v1})))\\ & +\mathbf{dv}\bullet(\mathbf{u2}\bullet(\mathbf{f}(\mathbf{v1} + \mathrm{dv2}\bullet\mathbf{u2} - \mathbf{f}(\mathbf{v1})))\\ & +\mathbf{dv}\bullet(\mathbf{u3}\bullet(\mathbf{f}(\mathbf{v1} + \mathrm{dv3}\bullet\mathbf{u31} - \mathbf{f}(\mathbf{v1})))\\ \rightarrow\ & \mathbf{dv}\bullet\mathbf{T}[\mathbf{D}[\mathbf{v1}]\bullet(\mathbf{f}(\mathbf{v});\mathbf{dv})]\end{aligned}$$

Consequently for functions continuously differentiable vectorially with a basic gradient at **v1**

$$\mathbf{f}(\mathbf{v1}+\mathbf{dv}) \approx \mathbf{f}(\mathbf{v1}) + \mathbf{dv}\bullet\mathbf{T}[\mathbf{D}[\mathbf{v1}]\bullet(\mathbf{f}(\mathbf{v});\mathbf{dv})].$$

If a function has a basic continuous gradient everywhere in **V3**, the function is said to have a basic continuous gradient **everywhere**."

Newton, interjecting: "This may not apply to other quadrants."

Breton: "Correct. If it does apply to all the other quadrants, the function is said to have a **simply continuous gradient**."

Newton, enjoying the questioning: "The gradient of a vectorial function is a matrix. What is its rank?"

Breton: "The answer depends on the function. A vector function may have a gradient unrelated to its neighboring value. Such gradients may have any rank."

Einstein: "But how about functions with basic gradients?"

Breton: "Then its rank depends on how **f(v1)** varies at **v1**. If the function varies only in one of the designated directions then its gradient has rank 1. Consequently it is possible for a gradient to be [**0**] for a function which does not vary in any of the designated directions, but nevertheless varies in some intermediate direction. If, however, a function is continuous in the positive quadrant, then a functional change in an intermediate direction is accompanied by a change in one of more of the designated directions according to the rank of the continuous gradient."

Einstein, pensively: "So with a changed orientation, a gradient may change not only its form but also its rank."

Breton: "Correct."

Newton, summarizing: "The promotion of a positive quadrant gradient into one simply continuous rests on two implications.
the function is continuously differentiable
the differentiability in the positive quadrant
extends to the other quadrants."

Breton: "The second implication is accommodated formally by allowing either one or more of the ***ui*** or the *dvi* to be negative."

Newton: "This must be related to directional gradients."

Breton: "Yes. Suppose the curve CQ contains **v1**. Let **f** be continuously differentiable vectorially at **v1**(q1) with a simply continuous gradient. Now for **dv** = **v1**(q1+dq) – **v1**(q1)

D[**v1**](**f**|CQ;dq) = D[**v**(q1)](**v**|CQ;dq)•**T**[**D**[**v1**]•(**f**(**v**);**dv**)].

Newton: "So this is a chain rule."

Breton: "The conditions for the validity of the above equation define the conditions for a valid chain rule in **V3**."

Einstein: "This equation carries some implications. Suppose **f**(**v**) is constant."

Breton: "Then

D[**v**(q1)](**v**|CQ;dq)•**T**[**D**[**v1**]•(**f**(**v**);**dv**)]= **0**.

Einstein: "And so D[**v**(q1)](**v**|CQ;dq) is in the null set of the gradient!"

Breton: "And D[**v**(q1)](**v**|CQ;dq) is the direction of the tangent line of **v**(q)."

Einstein, concluding: "So the tangent line of a curve lies in the null space of the gradient of a constant function defined on the curve!"

Breton: "There are other implications."

Newton, imitating Einstein: "How about gradients over sections?"

Sections

Breton: "The idea of a quadrant gradient can be further broadened by relaxing the requirement for orthogonality."

Einstein: "We need a definition."

Breton: "First let's examine sections in more detail. A diagram will be helpful."

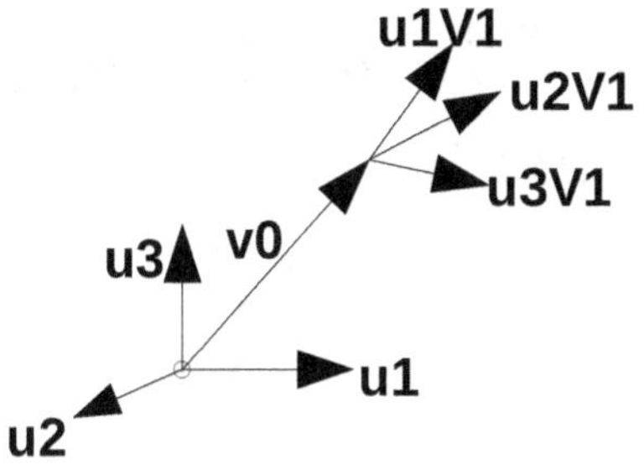

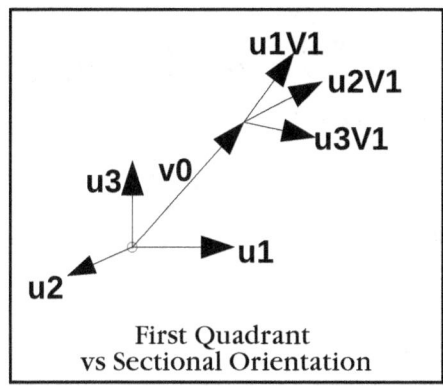

First Quadrant
vs Sectional Orientation

The origin is marked with an o which centers the origin's orientation. The diagram shows the three orthogonal vectors **u1**, **u2**, and **u3** for the origin's orientation. The section is also illustrated offset from the origin by the vector **v0**. The three defining vectors for the section, **u1V1**, **u2V1** and **u3V1** are shown. None of these vectors need lie in the planes defined by the origin's vectors."

Newton, after scribbling on a notepad: "These sections are hard to illustrate."

Breton: "So we must be that much more careful in defining them."

Newton: "There must be some relationship between the two."

Breton, accepting the challenge enthusiastically: "Let's find it. First consider the following transformation.

$$\mathbf{UV1} \equiv \mathbf{u1} * \mathbf{u1V1} + \mathbf{u2} * \mathbf{u2V1} + \mathbf{u3} * \mathbf{u3V1}$$

This matrix transforms a first quadrant vector into SECT(**0**,**u1V1**,**u2V1**,**u3V1**). What can be $\mathbf{UV1}^{-1}$, the inverse of UV1?"

Newton: "I can't imagine."

Breton, mirthfully: "Do you think our old friend, the scalar triple product could be useful here?"

Newton, seriously: "Perhaps. The ratio of respective volumes could form part of an answer."

Breton: "So let's form

$$\mathbf{u1} * \mathbf{u1V1} \cdot (\mathbf{u2V1} \wedge \mathbf{u3V1})/\mathrm{det}[\mathbf{UV1}].$$

Newton: "Which is just **u1**!"

Breton: "Then

$$\mathbf{u1} * (\mathbf{u1V1} \cdot (\mathbf{u2V1} \wedge \mathbf{u3V1})) * \mathbf{u1}/\mathrm{det}[\mathbf{UV1}] = \mathbf{u1} * \mathbf{u1}.$$

Newton, suddenly enlightened: "And likewise,

$$\mathbf{u2 \bullet (u2V1 \cdot (u3V1 \wedge u3V1)) \bullet u2} / \det[\mathbf{UV1}] = \mathbf{u2 \bullet u2}$$

and

$$\mathbf{u3 \bullet (u3V1 \cdot (u1V1 \wedge u2V1)) \bullet u3} / \det[\mathbf{UV1}] = \mathbf{u3 \bullet u3}.$$

Now we can put them all together to obtain

$$\begin{aligned}
&\mathbf{u1 \bullet (u1V1 \cdot (u2V1 \wedge u3V1)) \bullet u1} / \det[\mathbf{UV1}] \\
&\quad + \mathbf{u2 \bullet (u2V1 \cdot (u3V1 \wedge u1V1)) \bullet u2} / \det[\mathbf{UV1}] \\
&\quad + \mathbf{u3 \bullet (u3V1 \cdot (u1V1 \wedge u2V1)) \bullet u3} / \det[\mathbf{UV1}] \\
&\qquad = \mathbf{u \bullet u1 + u2 \bullet u2 + u3 \bullet u3} \\
&\qquad = \mathbf{I}.
\end{aligned}$$

Breton: "Now calculate

$$\begin{aligned}
&\mathbf{UV1} \cdot ((\mathbf{u2V1 \wedge u3V1) \bullet u1} \\
&\quad + (\mathbf{u3V1 \wedge u1V1) \bullet u2} \\
&\quad + (\mathbf{u1V1 \wedge u2V1) \bullet u3}) / \det[\mathbf{UV1}] \\
&\quad = \mathbf{u1 \bullet u1V1 + u2 \bullet u2V1 + u3 \bullet u3V1} \\
&\qquad ((\mathbf{u2V1 \wedge u3V1) \bullet u1} \\
&\qquad\quad + (\mathbf{u3V1 \wedge u1V1) \bullet u2} \\
&\qquad\quad + (\mathbf{u1V1 \wedge u2V1) \bullet u3}) / \det[\mathbf{UV1}] \\
&\quad = \mathbf{u1 \bullet u1V1} \cdot ((\mathbf{u2V1 \wedge u3V1) \bullet u1} \\
&\qquad\quad + (\mathbf{u3V1 \wedge u1V1) \bullet u2} \\
&\qquad\quad + (\mathbf{u1V1 \wedge u2V1) \bullet u3}) / \det[\mathbf{UV1}] \\
&\qquad + \mathbf{u2 \bullet u2V1} \cdot ((\mathbf{u2V1 \wedge u3V1) \bullet u1} \\
&\qquad\quad + (\mathbf{u3V1 \wedge u1V1) \bullet u2} \\
&\qquad\quad + (\mathbf{u1V1 \wedge u2V1) \bullet u3}) / \det[\mathbf{UV1}] \\
&\qquad + \mathbf{u3 \bullet u3V1} \cdot ((\mathbf{u2V1 \wedge u3V1) \bullet u1} \\
&\qquad\quad + (\mathbf{u3V1 \wedge u1V1) \bullet u2} \\
&\qquad\quad + (\mathbf{u1V1 \wedge u2V1) \bullet u3}) / \det[\mathbf{UV1}] \\
&\quad = \mathbf{u1 \bullet u1V1} \cdot ((\mathbf{u2V1 \wedge u3V1) \bullet u1} \\
&\qquad\quad + 0 \bullet \mathbf{u2} \\
&\qquad\quad + 0 \bullet \mathbf{u3}) / \det[\mathbf{UV1}] \\
&\qquad + \mathbf{u2 \bullet u2V1\ (u3V1 \wedge u1V1) \bullet u2} \\
&\qquad\quad + 0 \bullet \mathbf{u3} \\
&\qquad\quad + 0 \bullet \mathbf{u1}) / \det[\mathbf{UV1}] \\
&\qquad + \mathbf{u3 \bullet u3V1} \cdot (\mathbf{u1V1 \wedge u2V1) \bullet u3}) / \det[\mathbf{UV1}] \\
&\quad = \mathbf{u1} \bullet \det[\mathbf{UV1}] \bullet \mathbf{u1} / \det[\mathbf{UV1}] \\
&\qquad + \mathbf{u2} \bullet \det[\mathbf{UV1}] \bullet \mathbf{u2} / \det[\mathbf{UV1}] \\
&\qquad + \mathbf{u3} \bullet \det[\mathbf{UV1}] \bullet \mathbf{u3}) / \det[\mathbf{UV1}] \\
&\quad = \mathbf{u1 \bullet u1 + u2 \bullet u2 + u3 \bullet u3} \\
&\quad = \mathbf{I}.
\end{aligned}$$

So we have discovered

$$\begin{aligned}
\mathbf{UV1}^{-1} = {} & ((\mathbf{u2V1 \wedge u3V1) \bullet u1} \\
& + (\mathbf{u3V1 \wedge u1V1) \bullet u2} \\
& + (\mathbf{u1V1 \wedge u2V1) \bullet u3}) / \det[\mathbf{UV1}].
\end{aligned}$$

Einstein, concluding and leading: "So using **UV1** and $\mathbf{UV1}^{-1}$ we can map vectors between the first quadrant and SECT(**0,u1V1,u2V1,u3V1**). How do we get to SECT(**v0,u1V1,u2V1,u3V1**)?"

Breton: "By translating it. We simply add the vector **v0** to each element of SECT(**0,u1V1,u2V1,u3V1**).

Einstein, challenging: "Does the sectional notation also support an algebra?"

Breton: "Something similar to the algebra for the given orthogonal orientation. But before investigating that relationship, let us consider how a given vector is related in both orientations. Given vector **v1** in a section SECT(**0,u1V1,u2V1,u3V1**), let

$$\mathbf{v1} \equiv \text{v1V1}*\mathbf{u1V1} + \text{v2V1}*\mathbf{u2V1} + \text{v3V1}*\mathbf{u3V1}$$
$$\equiv \text{vV1}*\mathbf{uV1}.$$

The vector, **v1** may be transformed into a vector **v2** in the first quadrant.

$$\mathbf{v2} \equiv \mathbf{v1}\cdot\mathbf{UV1}^{-1}$$
$$\equiv \text{v21}*\mathbf{u1} + \text{v22}*\mathbf{u2} + \text{v23}*\mathbf{u3}$$
$$\equiv \text{v2}*\mathbf{uv}$$
$$\equiv \text{v2}*(\text{c1}*\mathbf{u1} + \text{c2}*\mathbf{u2} + \text{c3}*\mathbf{u3}).$$

In general, **v1** does not equal **v2**, nor does **uV1** = **uv**. Here is a somewhat confused illustration of what I mean.

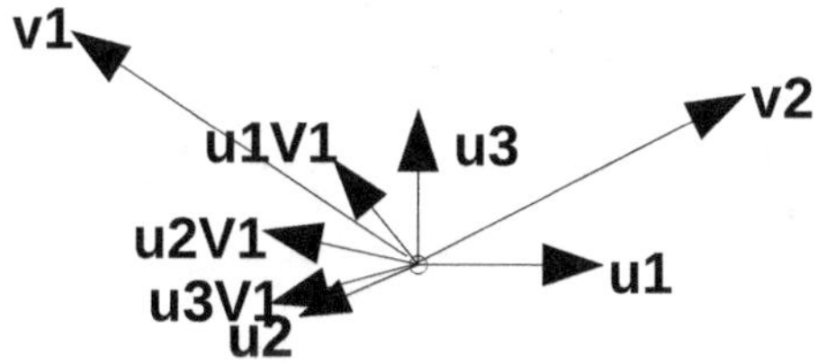

The orientation of the origin is given by the vectors **u1**, **u2**, and **u3** while the orientation of the section is given by **u1V1**, **u2V1**, and **u3V1**. The vector **v1** lies in the section; the transformed vector **v2** lies in the first quadrant.

The two vectors are related as

$$\mathbf{v1} = \mathbf{v2}\cdot\mathbf{UV1}$$
$$= \text{v2}*(\text{c1}*\mathbf{u1} + \text{c2}*\mathbf{u2} + \text{c3}*\mathbf{u3})$$
$$\cdot(\mathbf{u1}*\mathbf{u1V1} + \mathbf{u2}*\mathbf{u2V1} + \mathbf{u3}*\mathbf{u3V1})$$
$$= \text{v2}*(\text{c1}*\mathbf{u1V1} + \text{c2}*\mathbf{u2V1} + \text{c3}*\mathbf{u3V1})$$

so that

$\text{viV1} = \text{v2}*\text{ci}, \ \text{i}=1,2,3$

$\text{vV1} = \text{sqrt}(\mathbf{vV1}\cdot\mathbf{vV1})$

$= \text{sqrt}(\mathbf{v2}\cdot\mathbf{UV1}\cdot\mathbf{T}[\text{UV1}]\cdot\mathbf{v2})$

$$
\begin{aligned}
&= v2\ast\mathrm{sqrt}(c1^2\\
&\quad +c1\ast c2\ast\mathbf{u1V1}\bullet\mathbf{u2V1}\\
&\quad +c1\ast c3\ast\mathbf{u1V1}\bullet\mathbf{u3V1}\\
&\quad +c2\ast c1\ast\mathbf{u2V1}\bullet\mathbf{u1V1}\\
&\quad +c2^2\\
&\quad +c2\ast c3\ast\mathbf{u2V1}\bullet\mathbf{u3V1}\\
&\quad +c3\ast c1\ast\mathbf{u3V1}\bullet\mathbf{u1V1}\\
&\quad +c3\ast c2\ast\mathbf{u3V1}\bullet\mathbf{u2V1}\\
&\quad +c3^2)
\end{aligned}
$$

$$
\begin{aligned}
\mathbf{uV1} &= \mathbf{v1}/vV1\\
&= (v1V1\ast\mathbf{u1V1} + v2V1\ast\mathbf{u2V1} + v3V1\ast\mathbf{u3V1})/vV1.
\end{aligned}
$$

Einstein, returning to his original question: "How about the algebra, in particular how can two vectors be multiplied?"

Breton: "First let's do addition in SECT($\mathbf{v0},\mathbf{u1V1},\mathbf{u2V1},\mathbf{u3V1}$). Now let

$$
\begin{aligned}
\mathbf{v11} &\equiv \mathbf{v0} + v1V1\ast\mathbf{u11V1}\\
&\equiv \mathbf{v0} + v11V1\ast\mathbf{u1V1}\\
&\qquad + v12V1\ast\mathbf{u2V1}\\
&\qquad + v13V1\ast\mathbf{u3V1}\\
&= \mathbf{v0} + \mathbf{v21}\bullet\mathbf{UV1}
\end{aligned}
$$

and

$$
\begin{aligned}
\mathbf{v12} &\equiv \mathbf{v0} + v2V1\ast\mathbf{u12V1}\\
&\equiv \mathbf{v0} + v21V1\ast\mathbf{u1V1}\\
&\qquad + v22V1\ast\mathbf{u2V1}\\
&\qquad + v23V1\ast\mathbf{u3V1}\\
&= \mathbf{v0} + \mathbf{v22}\bullet\mathbf{UV1}
\end{aligned}
$$

where $\mathbf{u11V1}$ and $\mathbf{u12V1}$ are the directions of $\mathbf{v0} + \mathbf{v11}$ and $\mathbf{v0} + \mathbf{v12}$ respectfully.
Then

$$
\begin{aligned}
&\mathbf{v0} + \mathbf{v11} + \mathbf{v0} + \mathbf{v12}\\
&\quad= 2\ast\mathbf{v0}\\
&\qquad + v11V1\ast\mathbf{u1V1} + v12V1\ast\mathbf{u2V1} + v13V1\ast\mathbf{u3V1}\\
&\qquad + v21V1\ast\mathbf{u1V1} + v22V1\ast\mathbf{u2V1} + v23V1\ast\mathbf{u3V1}\\
&\quad= 2\ast\mathbf{v0}\\
&\qquad + (v11V1 + v21V1)\ast\mathbf{u1V1}\\
&\qquad + (v12V1 + v22V1)\ast\mathbf{u2V1}\\
&\qquad + (v13V1 + v23V1)\ast\mathbf{u3V1}\\
&\quad= 2\ast\mathbf{v0} + v1V1\ast\mathbf{u11V1} + v2V1\ast\mathbf{u12V1}.
\end{aligned}
$$

Einstein, victoriously: "In SECT($\mathbf{0},\mathbf{u1V1},\mathbf{u2V1},\mathbf{u3V1}$) where $\mathbf{v0} = \mathbf{0}$

$$
\begin{aligned}
\mathbf{v11} + \mathbf{v12} &= v11V1\ast\mathbf{u1V1} + v12V1\ast\mathbf{u2V1} + v13V1\ast\mathbf{u3V1}\\
&\quad + v21V1\ast\mathbf{u1V1} + v22V1\ast\mathbf{u2V1} + v23V1\ast\mathbf{u3V1}\\
&= v1V1\ast\mathbf{u11V1} + v2V1\ast\mathbf{u12V1}
\end{aligned}
$$

I conclude that sectorial addition differs in both sections."

Breton, taken aback: "And so it does. So let us first come to results for SECT($\mathbf{0},\mathbf{u1V1},\mathbf{u2V1},\mathbf{u3V1}$) before examining sections translated from the origin. Agreed?"

Newton, more interested in results: "Why not? We might just as well have considered the origin's orientation as being translated. Then too we would have had vectorial orientation modified. So let's consider first the orientations of the origin and a sector referred to the same zero of the set of vectors."

Sectional Algebra

Breton, plunging forward: "Then in terms of the given origin let the two sectional vectors be expressed in terms of the origin as

$$\mathbf{v11} \equiv v11*(c111*\mathbf{u1} + c112*\mathbf{u2} + c113*\mathbf{u3})$$
$$\mathbf{v12} \equiv v12*(c121*\mathbf{u1} + c122*\mathbf{u2} + c123*\mathbf{u3})$$

where the c's are directional cosines.
Then

$$\begin{aligned}\mathbf{v11} + \mathbf{v12} = {} & (v11*c111+ v12*c121)*\mathbf{u1} \\ & + (v11*c111+ v12*c122)*\mathbf{u2} \\ & + (v11*c113+ v12*c123)*\mathbf{u3}.\end{aligned}$$

Newton, advancing the argument: "Subtraction just replaces the positive with negative signs.

$$\begin{aligned}\mathbf{v11} - \mathbf{v12} = {} & (v11*c111- v12*c121)*\mathbf{u1} \\ & + (v11*c111- v12*c122)*\mathbf{u2} \\ & + (v11*c113- v12*c123)*\mathbf{u3}.\end{aligned}$$

Einstein, still looking for obstructions: "That may be different for translated sections."

Newton, in riposte: "Let's keep focused on sections referred to the origin."

Einstein, questioning: "How does this addition relate to the addition of the transformed vector of the given orthogonal orientation?"

Breton: "Simple enough. Let **v21** and **v22** be the vectors transformed from the section into the first quadrant. Then

$$\begin{aligned}\mathbf{v21} + \mathbf{v22} & = \mathbf{v11} \bullet \mathbf{UV1}^{-1} + \mathbf{v12} \bullet \mathbf{UV1}^{-1} \\ & = (\mathbf{v11} + \mathbf{v12}) \bullet \mathbf{UV1}^{-1}\end{aligned}$$

Subtraction follows similarly."

Einstein: "Let's move on to the multiplications."

Breton: "Then as before for

$$\mathbf{v11} = v11*(c111*\mathbf{u1} + c112*\mathbf{u2} + c113*\mathbf{u3})$$
$$\mathbf{v12} = v12*(c121*\mathbf{u1} + c122*\mathbf{u2} + c123*\mathbf{u3})$$
$$\begin{aligned}\mathbf{v11}*\mathbf{v12} = v11*v12*(& c111*c121*\mathbf{u1}*\mathbf{u1} \\ & + c111*c122*\mathbf{u1}*\mathbf{u2} \\ & + c111*c123*\mathbf{u1}*\mathbf{u3} \\ & + c112*c121*\mathbf{u2}*\mathbf{u1} \\ & + c112*c122*\mathbf{u2}*\mathbf{u2} \\ & + c112*c123*\mathbf{u2}*\mathbf{u3}\end{aligned}$$

$+\ c113*c121*\mathbf{u3}*\mathbf{u1}$
$+\ c113*c122*\mathbf{u3}*\mathbf{u2}$
$+\ c113*c123*\mathbf{u3}*\mathbf{u3}$
$\mathbf{v11}\cdot\mathbf{v12} = v11*v12*(c111*c121+ c112*c122 + c113*c123$
$\mathbf{v11}\wedge\mathbf{v12} = ((c112*c123-c113*c122)*\mathbf{u1}$
$+ (c113*c121-c111*c123)*\mathbf{u2}$
$+ (c111*c122-c112*c121)*\mathbf{u3}.$

Einstein, pushing his agenda: "This result is expressed in terms of the origin's orientation. What can be the result of the multiplications expressed in terms of the section's orientation? Give us the results in terms of the orientation of the sector!

Breton: "From the definitions then

$\mathbf{v11} \equiv v111V1*\mathbf{u1V1} + v112V1*\mathbf{u2V1} + v113V1*\mathbf{u3V1}$
$\equiv v11V1*\mathbf{u11V1}$
$\mathbf{v12} \equiv v121V1*\mathbf{u1V1} + v122V1*\mathbf{u2V1} + v123V1*\mathbf{u3V1}$
$\equiv v12V1*\mathbf{u12V1}$
$\mathbf{v11}*\mathbf{v12} = v11V1*u11V1*\mathbf{u11V1}*\mathbf{u12V1}$
$\mathbf{v11}\cdot\mathbf{v12} = v11V1*u11V1*\mathbf{u11V1}\cdot\mathbf{u12V1}$
$\mathbf{v11}\wedge\mathbf{v12} = v11V1*u11V1*\mathbf{u11V1}\wedge\mathbf{u12V1}.$

Einstein, stubbornly insisting on leading: "Your result is given in terms of directions. Show us the result in terms of the sectional orienting vectors."

Breton: "You like to do bookkeeping? So here are the results in terms of sectional orientation.

$\mathbf{v11}*\mathbf{v12} = (v111V1*\mathbf{u1V1} + v112V1*\mathbf{u2V1} + v113V1*\mathbf{u3V1})$
$*(v121V1*\mathbf{u1V1} + v122V1*\mathbf{u2V1} + v123V1*\mathbf{u3V1})$
$= v111V1*v121V1*\mathbf{u1V1}*\mathbf{u1V1}$
$+ v111V1*v122V1*\mathbf{u1V1}*\mathbf{u2V1}$
$+ v111V1*v123V1*\mathbf{u1V1}*\mathbf{u3V1}$
$+ v112V1*v121V1*\mathbf{u2V1}*\mathbf{u1V1}$
$+ v112V1*v122V1*\mathbf{u2V1}*\mathbf{u2V1}$
$+ v112V1*v123V1*\mathbf{u2V1}*\mathbf{u3V1}$
$+ v113V1*v121V1*\mathbf{u3V1}*\mathbf{u1V1}$
$+ v113V1*v122V1*\mathbf{u3V1}*\mathbf{u2V1}$
$+ v113V1*v123V1*\mathbf{u3V1}*\mathbf{u3V1}.$

Newton, with amused toleration: "And so

$$
\begin{aligned}
\mathbf{v11}\bullet\mathbf{v12} = {} & v111V1*v121V1*\mathbf{u1V1}\bullet\mathbf{u1V1} \\
& + v111V1*v122V1*\mathbf{u1V1}\bullet\mathbf{u2V1} \\
& + v111V1*v123V1*\mathbf{u1V1}\bullet\mathbf{u3V1} \\
& + v112V1*v121V1*\mathbf{u2V1}\bullet\mathbf{u1V1} \\
& + v112V1*v122V1*\mathbf{u2V1}\bullet\mathbf{u2V1} \\
& + v112V1*v123V1*\mathbf{u2V1}\bullet\mathbf{u3V1} \\
& + v113V1*v121V1*\mathbf{u3V1}\bullet\mathbf{u1V1} \\
& + v113V1*v122V1*\mathbf{u3V1}\bullet\mathbf{u2V1} \\
& + v113V1*v123V1*\mathbf{u3V1}\bullet\mathbf{u3V1}.
\end{aligned}
$$

Einstein, mimicking: "And so

$$
\begin{aligned}
\mathbf{v11}\wedge\mathbf{v12} = {} & v111V1*v121V1*\mathbf{u1V1}\wedge\mathbf{u1V1} \\
& + v111V1*v122V1*\mathbf{u1V1}\wedge\mathbf{u2V1} \\
& + v111V1*v123V1*\mathbf{u1V1}\wedge\mathbf{u3V1} \\
& + v112V1*v121V1*\mathbf{u2V1}\wedge\mathbf{u1V1} \\
& + v112V1*v122V1*\mathbf{u2V1}\wedge\mathbf{u2V1} \\
& + v112V1*v123V1*\mathbf{u2V1}\wedge\mathbf{u3V1} \\
& + v113V1*v121V1*\mathbf{u3V1}\wedge\mathbf{u1V1} \\
& + v113V1*v122V1*\mathbf{u3V1}\wedge\mathbf{u2V1} \\
& + v113V1*v123V1*\mathbf{u3V1}\wedge\mathbf{u3V1}.
\end{aligned}
$$

With this Newton pointed his aquiline nose at Einstein with an amused smile which neatly expressed his amusement. Breton chuckled. Einstein continued unabashedly.

Einstein: "How about the relationship between the vectors transferred into the first quadrant?"

Breton: "Since the first quadrant vectors can be effected by a transformation from the sector vectors, there should indeed be a relationship between the two."

Newton, turning to Breton: "Let's find it."

Breton: "Let

$$\mathbf{v21} \equiv \mathbf{v11}\bullet\mathbf{UV1}^{-1}$$
$$\mathbf{v22} \equiv \mathbf{v12}\bullet\mathbf{UV1}^{-1}$$

Then

$$\mathbf{v21}*\mathbf{v22} = [\mathbf{UV1}^{-1}]\bullet\mathbf{v11}*\mathbf{v12}\bullet\mathbf{UV1}^{-1}$$
$$\mathbf{v21}\bullet\mathbf{v22} = tr[[\mathbf{UV1}^{-1}]\bullet\mathbf{v11}*\mathbf{v12}\bullet\mathbf{UV1}^{-1}]$$
$$\mathbf{v21}\wedge\mathbf{v22} = \mathbf{c}[[\mathbf{UV1}^{-1}]\bullet\mathbf{v11}*\mathbf{v12}\bullet\mathbf{UV1}^{-1}]$$

where tr is the trace of a matrix and **c** is the curl matrix operator."

Einstein: "Well now, just go on to show products of a vector in the section with the accompanying vector in the first quadrant."

Breton: "Do you think the bookkeeping will be less intense?"

Newton: "The results can be expressed either in the first quadrant orientation or the sectional orientation."

Breton: "So let's start with the origin's orientation."
v11 = v21•UV1
v21 = v2*uv
so
v11*v21 =T[UV1]•v21*v21
= u1V1*u1+u2V1*u2+u3V1*u3
• (v21*u1 + v22*u2 + v23*u3)
*(v21*u1 + v22*u2 + v23*u3)
= u1V1*u1+u2V1*u2+u3V1*u3
•(v21*v21*u1*u1 + v21*v22*u1*u2 + v21*v23*u1*u3
+ v22*v21*u2*u1 + v22*v22*u2*u2 + v22*v23*u2*u3
+ v23*v21*u3*u1 + v23*v22*u3*u2 + v23*v23*u3*u3)
= v21*v21*u1V1*u1
+ v21*v22*u1V1*u2
+ v21*v23*u1V1*u3
+ v22*v21*u2V1*u1
+ v22*v22*u2V1*u2
+ v22*v23*u2V1*u3
+ v23*v21*u3V1*u1
+ v23*v22*u3V1*u2
+ v23*v23*u3V1*u3
That settles the result for the origin's orientation."

Einstein: "And for the sector's orientation?"

Breton: "Then
v11*v21 = v11*v11•$[UV1]^{-1}$
and as we have seen
$UV1^{-1}$ = ((u2V1∧u3V1)*u1
+ (u3V1∧u1V1)*u2
+ (u1V1∧u2V1)*u3)/det[UV1]
so that
v11•$[UV1]^{-1}$ = v111V1*u1V1 + v112V1*u2V1 + v113V1*u3V1
•((u2V1∧u3V1)*u1
+ (u3V1∧u1V1)*u2
+ (u1V1∧u2V1)*u3)/det[UV1]
= v111V1*u1 + v112V1*u2 + v113V1*u3
and
v11*v21 =(v111V1*u1V1 + v112V1*u2V1 + v113V1*u3V1)
*(v111V1*u1 + v112V1*u2 + v113V1*u3)
= (v111V1*v111V1*u1V1*u1
+v111V1*v112V1*u1V1*u2
+v111V1*v113V1*u1V1*u3
+v112V1*v111V1*u2V1*u1
+v112V1*v112V1*u2V1*u2
+v112V1*v113V1*u2V1*u3
+v113V1*v111V1*u3V1*u1
+v113V1*v112V1*u3V1*u2
+v113V1*v113V1*u3V1*u3

That settles the result for the section's orientation."

Einstein: "Then do divergence and curls follow the same path?"

Breton: "Yes."

Einstein, following a now familiar path: "How about triple products?"

Breton: "In this regard let me prove a significant vectorial relationship which we did not prove in tp1.2. For any three vectors

$$(\mathbf{v1}\wedge\mathbf{v2})\ast\mathbf{v3} + (\mathbf{v3}\wedge\mathbf{v1})\ast\mathbf{v2} + (\mathbf{v2}\wedge\mathbf{v3})\ast\mathbf{v1} = ((\mathbf{v1}\wedge\mathbf{v2})\bullet\mathbf{v3})\ast I.$$

Einstein, happily welcoming the diversion: "Show us."

Breton: "Start with $\mathbf{v1}\wedge\mathbf{v2})\ast\mathbf{v3}$. For

$$\mathbf{v1} \equiv v11\ast\mathbf{u1} + v12\ast\mathbf{u2} + v13\ast\mathbf{u3}$$
$$\mathbf{v2} \equiv v21\ast\mathbf{u1} + v22\ast\mathbf{u2} + v23\ast\mathbf{u3}$$
$$\mathbf{v3} \equiv v31\ast\mathbf{u1} + v32\ast\mathbf{u2} + v33\ast\mathbf{u3}$$
$$\begin{aligned}\mathbf{v1}\wedge\mathbf{v2} = &(v13\ast v22 - v12\ast v23)\ast\mathbf{u1}\\ &+ (v11\ast v23 - v13\ast v21)\ast\mathbf{u2}\\ &+ (v12\ast v21 - v11\ast v22)\ast\mathbf{u3}\end{aligned}$$

so

$$\begin{aligned}(\mathbf{v1}\wedge\mathbf{v2})\ast\mathbf{v3} = &(v13\ast v22 - v12\ast v23)\ast\mathbf{u1}\\ &+ (v11\ast v23 - v13\ast v21)\ast\mathbf{u2}\\ &+ (v12\ast v21 - v11\ast v22)\ast\mathbf{u3})\\ &\qquad\ast(v31\ast\mathbf{u1} + v32\ast\mathbf{u2} + v33\ast\mathbf{u3})\\ = &(v13\ast v22 - v12\ast v23)\ast v31\ast\mathbf{u1}\ast\mathbf{u1}\\ &+ (v13\ast v22 - v12\ast v23)\ast v32\ast\mathbf{u1}\ast\mathbf{u2}\\ &+ (v13\ast v22 - v12\ast v23)\ast v33\ast\mathbf{u1}\ast\mathbf{u3}\\ &+ (v11\ast v23 - v13\ast v21)\ast v31\ast\mathbf{u2}\ast\mathbf{u1}\\ &+ (v11\ast v23 - v13\ast v21)\ast v32\ast\mathbf{u2}\ast\mathbf{u2}\\ &+ (v11\ast v23 - v13\ast v21)\ast v33\ast\mathbf{u2}\ast\mathbf{u3}\\ &+ (v12\ast v21 - v11\ast v22)\ast v31\ast\mathbf{u3}\ast\mathbf{u1}\\ &+ (v12\ast v21 - v11\ast v22)\ast v32\ast\mathbf{u3}\ast\mathbf{u2}\\ &+ (v12\ast v21 - v11\ast v22)\ast v33\ast\mathbf{u3}\ast\mathbf{u3}\end{aligned}$$

Likewise

$$\begin{aligned}(\mathbf{v2}\wedge\mathbf{v3})\ast\mathbf{v1} = &(v23\ast v32 - v22\ast v33)\ast\mathbf{u1}\\ &+ (v21\ast v33 - v23\ast v31)\ast\mathbf{u2}\\ &+ (v22\ast v31 - v21\ast v32)\ast\mathbf{u3})\\ &\qquad\ast(v11\ast\mathbf{u1} + v12\ast\mathbf{u2} + v13\ast\mathbf{u3})\\ = &(v23\ast v32 - v22\ast v33)\ast v11\ast\mathbf{u1}\ast\mathbf{u1}\\ &+ (v23\ast v32 - v22\ast v33)\ast v12\ast\mathbf{u1}\ast\mathbf{u2}\\ &+ (v23\ast v32 - v22\ast v33)\ast v13\ast\mathbf{u1}\ast\mathbf{u3}\\ &+ (v21\ast v33 - v23\ast v31)\ast v11\ast\mathbf{u2}\ast\mathbf{u1}\\ &+ (v21\ast v33 - v23\ast v31)\ast v12\ast\mathbf{u2}\ast\mathbf{u2}\\ &+ (v21\ast v33 - v23\ast v31)\ast v13\ast\mathbf{u2}\ast\mathbf{u3}\\ &+ (v22\ast v31 - v21\ast v32)\ast v11\ast\mathbf{u3}\ast\mathbf{u1}\\ &+ (v22\ast v31 - v21\ast v32)\ast v12\ast\mathbf{u3}\ast\mathbf{u2}\\ &+ (v22\ast v31 - v21\ast v32)\ast v13\ast\mathbf{u3}\ast\mathbf{u3}\end{aligned}$$

and

$$
\begin{aligned}
(\mathbf{v3}\wedge\mathbf{v1})\ast\mathbf{v2} = {} & (v33\ast v12 - v32\ast v13)\ast\mathbf{u1} \\
& + (v31\ast v13 - v33\ast v11)\ast\mathbf{u2} \\
& + (v32\ast v11 - v31\ast v12)\ast\mathbf{u3}) \\
& \qquad \ast(v21\ast\mathbf{u1} + v22\ast\mathbf{u2} + v23\ast\mathbf{u3}) \\
= {} & (v33\ast v12 - v32\ast v13)\ast v21\ast\mathbf{u1}\ast\mathbf{u1} \\
& + (v33\ast v12 - v32\ast v13)\ast v22\ast\mathbf{u1}\ast\mathbf{u2} \\
& + (v33\ast v12 - v32\ast v13)\ast v23\ast\mathbf{u1}\ast\mathbf{u3} \\
& + (v31\ast v13 - v33\ast v11)\ast v21\ast\mathbf{u2}\ast\mathbf{u1} \\
& + (v31\ast v13 - v33\ast v11)\ast v22\ast\mathbf{u2}\ast\mathbf{u2} \\
& + (v31\ast v13 - v33\ast v11)\ast v23\ast\mathbf{u2}\ast\mathbf{u3} \\
& + (v32\ast v11 - v31\ast v12)\ast v21\ast\mathbf{u3}\ast\mathbf{u1} \\
& + (v32\ast v11 - v31\ast v12)\ast v22\ast\mathbf{u3}\ast\mathbf{u2} \\
& + (v32\ast v11 - v31\ast v12)\ast v23\ast\mathbf{u3}\ast\mathbf{u3}.
\end{aligned}
$$

Newton, who had been following closely: "Let me put them together.

$$
\begin{aligned}
& (\mathbf{v1}\wedge\mathbf{v2})\ast\mathbf{v3} + (\mathbf{v2}\wedge\mathbf{v3})\ast\mathbf{v1} + (\mathbf{v3}\wedge\mathbf{v1})\ast\mathbf{v2} \\
& = (v13\ast v22 - v12\ast v23)\ast v31\ast\mathbf{u1}\ast\mathbf{u1} \\
& + (v13\ast v22 - v12\ast v23)\ast v32\ast\mathbf{u1}\ast\mathbf{u2} \\
& + (v13\ast v22 - v12\ast v23)\ast v33\ast\mathbf{u1}\ast\mathbf{u3} \\
& + (v11\ast v23 - v13\ast v21)\ast v31\ast\mathbf{u2}\ast\mathbf{u1} \\
& + (v11\ast v23 - v13\ast v21)\ast v32\ast\mathbf{u2}\ast\mathbf{u2} \\
& + (v11\ast v23 - v13\ast v21)\ast v33\ast\mathbf{u2}\ast\mathbf{u3} \\
& + (v12\ast v21 - v11\ast v22)\ast v31\ast\mathbf{u3}\ast\mathbf{u1} \\
& + (v12\ast v21 - v11\ast v22)\ast v32\ast\mathbf{u3}\ast\mathbf{u2} \\
& + (v12\ast v21 - v11\ast v22)\ast v33\ast\mathbf{u3}\ast\mathbf{u3} \\
& + (v23\ast v32 - v22\ast v33)\ast v11\ast\mathbf{u1}\ast\mathbf{u1} \\
& + (v23\ast v32 - v22\ast v33)\ast v12\ast\mathbf{u1}\ast\mathbf{u2} \\
& + (v23\ast v32 - v22\ast v33)\ast v13\ast\mathbf{u1}\ast\mathbf{u3} \\
& + (v21\ast v33 - v23\ast v31)\ast v11\ast\mathbf{u2}\ast\mathbf{u1} \\
& + (v21\ast v33 - v23\ast v31)\ast v12\ast\mathbf{u2}\ast\mathbf{u2} \\
& + (v21\ast v33 - v23\ast v31)\ast v13\ast\mathbf{u2}\ast\mathbf{u3} \\
& + (v22\ast v31 - v21\ast v32)\ast v11\ast\mathbf{u3}\ast\mathbf{u1} \\
& + (v22\ast v31 - v21\ast v32)\ast v12\ast\mathbf{u3}\ast\mathbf{u2} \\
& + (v22\ast v31 - v21\ast v32)\ast v13\ast\mathbf{u3}\ast\mathbf{u3} \\
& + (v33\ast v12 - v32\ast v13)\ast v21\ast\mathbf{u1}\ast\mathbf{u1} \\
& + (v33\ast v12 - v32\ast v13)\ast v22\ast\mathbf{u1}\ast\mathbf{u2} \\
& + (v33\ast v12 - v32\ast v13)\ast v23\ast\mathbf{u1}\ast\mathbf{u3} \\
& + (v31\ast v13 - v33\ast v11)\ast v21\ast\mathbf{u2}\ast\mathbf{u1} \\
& + (v31\ast v13 - v33\ast v11)\ast v22\ast\mathbf{u2}\ast\mathbf{u2} \\
& + (v31\ast v13 - v33\ast v11)\ast v23\ast\mathbf{u2}\ast\mathbf{u3} \\
& + (v32\ast v11 - v31\ast v12)\ast v21\ast\mathbf{u3}\ast\mathbf{u1} \\
& + (v32\ast v11 - v31\ast v12)\ast v22\ast\mathbf{u3}\ast\mathbf{u2} \\
& + (v32\ast v11 - v31\ast v12)\ast v23\ast\mathbf{u3}\ast\mathbf{u3}.
\end{aligned}
$$

Einstein: "Assemble the various components so we can see whether Breton's assertion is correct"

Newton: "For

u1*u1

(v13*v22 - v12*v23)*v31
+ (v23*v32 - v22*v33)*v11
+ (v33*v12 - v32*v13)*v21
= v13*v22*v31 - v12*v23*v31
+ v23*v32*v11 - v22*v33*v11
+ v33*v12*v21 - v32*v13*v21

u1*u2

(v13*v22 - v12*v23)*v32
+ (v23*v32 - v22*v33)*v12
+ (v33*v12 - v32*v13)*v22
= v13*v22*v32 - v12*v23*v32
+ v23*v32*v12 -v22*v33*v12
+ v33*v12*v22 - v32*v13*v22
= 0

u1*u3

(v13*v22 - v12*v23)*v33
+ (v23*v32 - v22*v33)*v13
+ (v33*v12 - v32*v13)*v23
= v13*v22*v33 - v12*v23*v33
+ v23*v32*v13 - v22*v33*v13
+ v33*v12*v23 - v32*v13*v23
= 0

u2*u1

(v11*v23 - v13*v21)*v31
+ (v21*v33 - v23*v31)*v11
+ (v31*v13 - v33*v11)*v21
= v11*v23*v31 - v13*v21*v31
+(v21*v33*v11 - v23*v31*v11
+(v31*v13*v21 - v33*v11*v21
= 0

u2*u2

(v11*v23 - v13*v21)*v32
+ (v21*v33 - v23*v31)*v12
+ (v31*v13 - v33*v11)*v22
= v11*v23*v32 - v13*v21*v32
+ v21*v33*v12 - v23*v31*v12
+ v31*v13*v22 - v33*v11*v22

u2*u3

(v11*v23 - v13*v21)*v33
+ (v21*v33 - v23*v31)*v13
+ (v31*v13 - v33*v11)*v23
= (v11*v23*v33 - v13*v21*v33
+ (v21*v33*v13 - v23*v31*v13
+ (v31*v13*v23 - v33*v11*v23
= 0

u3*u1

(v12*v21 – v11*v22)*v31
+ (v22*v31 – v21*v32)*v11
+ (v31*v13 – v33*v11)*v21
= v12*v21*v31 – v11*v22*v31
+ v22*v31*v11 – v21*v32*v11
+ v32*v11*v21 – v31*v12*v21
= 0

u3*u2

(v12*v21 – v11*v22)*v32
+ (v22*v31 – v21*v32)*v12
+ (v32*v11 – v31*v12)*v22
= v12*v21*v32 – v11*v22*v32
+ v22*v31*v12 – v21*v32*v12
+ v32*v11*v22 – v31*v12*v22
= 0

u3*u3

(v12*v21 – v11*v22)*v33
+ (v22*v31 – v21*v32)*v13
+ (v32*v11 – v31*v12)*v23
= v12*v21*v33 – v11*v22*v33
+ v22*v31*v13 – v21*v32*v13
+ v32*v11*v23 – v31*v12*v23.

Einstein: "Amazing. All the non-diagonal elements equal 0. Let me put the diagonal elements together."
For

u1*u1

v13*v22*v31 – v12*v23*v31
+ v23*v32*v11 – v22*v33*v11
+ v33*v12*v21 – v32*v13*v21

u2*u2

v11*v23*v32 – v13*v21*v32
+ (v21*v33*v12 – v23*v31*v12
+ (v31*v13*v22 – v33*v11*v22

u3*u3

v12*v21*v33 – v11*v22*v33
+ v22*v31*v13 – v21*v32*v13
+ v32*v11*v23 – v31*v12*v23

What is all this?"

Breton: "Look closely. All of them are equal although the factors are differently arranged. Now please calculate $((\mathbf{v1} \wedge \mathbf{v2}) \cdot \mathbf{v3})$."

Einstein: "All right

$$
\begin{aligned}
(\mathbf{v1}\wedge\mathbf{v2})\bullet\mathbf{v3}) &= (v13\ast v22 - v12\ast v23)\ast\mathbf{u1} \\
&\quad + (v11\ast v23 - v13\ast v21)\ast\mathbf{u2} \\
&\quad + (v12\ast v21 - v11\ast v22)\ast\mathbf{u3} \\
&\qquad \bullet(v31\ast\mathbf{u1} + v32\ast\mathbf{u2} + v33\ast\mathbf{u3}) \\
&= (v13\ast v22 - v12\ast v23)\ast v31 \\
&\quad + (v11\ast v23 - v13\ast v21)\ast v32 \\
&\quad + (v12\ast v21 - v11\ast v22)\ast v33) \\
&= v13\ast v22\ast v31 - v12\ast v23\ast v31 \\
&\quad + v11\ast v23\ast v32 - v13\ast v21\ast v32 \\
&\quad + v12\ast v21\ast v33 - v11\ast v22\ast v33)
\end{aligned}
$$

The factors are all the same."

Breton: "And so after no little algebraic calculations we can assert

$$(\mathbf{v1}\wedge\mathbf{v2})\ast\mathbf{v3} + (\mathbf{v3}\wedge\mathbf{v1})\ast\mathbf{v2} + (\mathbf{v2}\wedge\mathbf{v3})\ast\mathbf{v1} = ((\mathbf{v1}\wedge\mathbf{v2})\bullet\mathbf{v3})\ast\mathbf{I}.$$

Newton: "Our triple scalar product has become more important than I first realized. Now we see it in another light."

Einstein, recognizing the diversion as not diverting: "So how does this result help with triple products?"

Breton: "Notice $((\mathbf{v1}\wedge\mathbf{v2})\bullet\mathbf{v3})$ equals $\det[\mathbf{UV1}]$. We can start with

$$
\begin{aligned}
\det[\mathbf{UV1}]\ast\mathbf{I} &= (\mathbf{u2V1}\wedge\mathbf{u3V1})\ast\mathbf{u1V1} \\
&\quad + (\mathbf{u3V1}\wedge\mathbf{u1V1})\ast\mathbf{u2V1} \\
&\quad + (\mathbf{u1V1}\wedge\mathbf{u2V1})\ast\mathbf{u3V1}
\end{aligned}
$$

so

$$
\begin{aligned}
\mathrm{tr}[\det[\mathbf{UV1}]\ast\mathbf{I}] &= (\mathbf{u2V1}\wedge\mathbf{u3V1})\bullet\mathbf{u1V1} \\
&\quad + (\mathbf{u3V1}\wedge\mathbf{u1V1})\bullet\mathbf{u2V1} \\
&\quad + (\mathbf{u1V1}\wedge\mathbf{u2V1})\bullet\mathbf{u3V1} \\
&= 3\ast\det[\mathbf{UV1}]
\end{aligned}
$$

and remembering $\mathbf{c}[\mathbf{v1}\ast\mathbf{v2}] = \mathbf{v1}\wedge\mathbf{v2}$

$$
\begin{aligned}
\mathbf{c}(\det[\mathbf{UV1}]\ast\mathbf{I}) &= (\mathbf{u2V1}\wedge\mathbf{u3V1})\wedge\mathbf{u1V1} \\
&\quad + (\mathbf{u3V1}\wedge\mathbf{u1V1})\wedge\mathbf{u2V1} \\
&\quad + (\mathbf{u1V1}\wedge\mathbf{u2V1})\wedge\mathbf{u3V1} \\
&= \mathbf{0}.
\end{aligned}
$$

Newton: "Many are these equations. Let me put them into a table."

After a few minutes Newton produced the following table which he handed to his friends.

Item	section's orientation	origin's orientation
v11	v11V1*u1V1 + v12V1*u2V1 + v13V1*u3V1	v11*c111*u1 + v11*c112*u2 + v11*c113*u3
v12	v21V1*u1V1 + v22V1*u2V1 + v23V1*u3V1	v12*c121*u1 + v12*c122*u2 + v12*c123*u3
v11 + **v12**	v1V1*u11V1 +v2V1*u12V1 = (v11V1 + v21V1) *u1V1 + (v12V1 + v22V1) *u2V1 + (v13V1 + v23V1) *u3V1	(v11*c111 + v12*c121)*u1 + (v11*c111 + v12*c122)*u2 + (v11*c113 + v12*c123)*u3
v11 - v12	v1V1*u11V1 -v2V1*u12V1 = (v11V1 - v21V1) *u1V1 + (v12V1 - v22V1) *u2V1 + (v13V1- v23V1) *u3V1	(v11*c111 - v12*c121)*u1 + (v11*c111 -v12*c122)*u2 + (v11*c113 - v12*c123)*u3
v11*v12	v111V1*v121V1 *u1V1*u1V1 + v111V1*v122V1 *u1V1*u2V1 + v111V1*v123V1 *u1V1*u3V1 + v112V1*v121V1 *u2V1*u1V1 + v112V1*v122V1 *u2V1*u2V1 + v112V1*v123V1 *u2V1*u3V1 + v113V1*v121V1 *u3V1*u1V1 + v113V1*v122V1 *u3V1*u2V1 + v113V1*v123V1 *u3V1*u3V1	v11*v12 *(c111*c121*u1*u1 + c111*c122*u1*u2 + c111*c123*u1*u3 + c112*c121*u2*u1 + c112*c122*u2*u2 + c112*c123*u2*u3 + c113*c121*u3*u1 + c113*c122*u3*u2 + c113*c123*u3*u3

v11•v12	v111V1*v121V1 ***u1V1•u1V1** + v111V1*v122V1 ***u1V1•u2V1** + v111V1*v123V1 ***u1V1•u3V1** + v112V1*v121V1 ***u2V1•u1V1** + v112V1*v122V1 ***u2V1•u2V1** + v112V1*v123V1 ***u2V1•u3V1** + v113V1*v121V1 ***u3V1•u1V1** + v113V1*v122V1 ***u3V1•u2V1** + v113V1*v123V1 ***u3V1•u3V1**	v11*v12 *(c111*c121 + c112*c122 + c113*c123)
v11∧v2	v111V1*v121V1 ***u1V1∧u1V1** + v111V1*v122V1 ***u1V1∧u2V1** + v111V1*v123V1 ***u1V1∧u3V1** + v112V1*v121V1 ***u2V1∧u1V1** + v112V1*v122V1 ***u2V1∧u2V1** + v112V1*v123V1 ***u2V1∧u3V1** + v113V1*v121V1 ***u3V1∧u1V1** + v113V1*v122V1 ***u3V1∧u2V1** + v113V1*v123V1 ***u3V1∧u3V1**	((c112*c123 −c113*c122)***u1** + (c113*c121 −c111*c123)***u2** + (c111*c122 −c112*c121)***u3**
v21*v22	[**UV1**$^{-1}$] **•v11*v12•UV1**$^{-1}$	
v21•v22	tr[[**UV1**$^{-1}$] **•v11*v12•UV1**$^{-1}$]	
v21∧v22	**c**[[**UV1**$^{-1}$] **•v11*v12•UV1**$^{-1}$]	

v11*v21	(v111V1*v111V1 *u1V1*u1 +v111V1*v112V1 *u1V1*u2 +v111V1*v113V1 *u1V1*u3 +v112V1*v111V1 *u2V1*u1 +v112V1*v112V1 *u2V1*u2 +v112V1*v113V1 *u2V1*u3 +v113V1*v111V1 *u3V1*u1 +v113V1*v112V1 *u3V1*u2 +v113V1*v113V1 *u3V1*u3	v21*v21*u1V1*u1 + v21*v22*u1V1*u2 + v21*v23*u1V1*u3 + v22*v21*u2V1*u1 + v22*v22*u2V1*u2 + v22*v23*u2V1*u3 + v23*v21*u3V1*u1 + v23*v22*u3V1*u2 + v23*v23*u3V1*u3
(u2V1∧u3V1) •u1V1	det[UV1]	
(u2V1 ∧u3V1) ∧u1V1 + (u3V1 ∧u1V1) ∧u2V1 + (u1V1 ∧u2V1) ∧u3V1	0	

Einstein, noting: "The table only applies to
SECT(**0,u1V1**,**u2V1**,**u3V1**)."

Sectors displaced from the origin

Breton: "True enough. For the sector SECT(**v0,u1V1**,**u2V1**,**u3V1**) the results would have to include the influence of the displacement vector **v0**. In particular the matrix **UV1** no longer maps a first quadrant vector into the displaced sector. Two vectors, **v1** in the section and its image, **v2**, in the first quadrant are related as

v1 = **v0** + **V2**•**UV1**

The vector **v1** is now written in terms of the sectional orientation as

v1 ≡ **v0** + v1V1*u1V1 + v2V1*u2V1 + v3V1*u3V1
= **v0** + v1V1*u1V1 + v2V1*u2V1 + v3V1*u3V1
≡ **v0** + vV1*uV1.

Einstein: "How is **v1** transformed into a vector in the first quadrant?"

Breton: "Since
v1 − v0 = v1V1∗**u1V1** + v2V1∗**u2V1** + v3V1∗**u3V1**
v2 = (**v1 − v0**)•**UV1**$^{-1}$
≡ v21∗**u1** + v22∗**u2** + v23∗**u3**
≡ v2∗**uv**
≡ v2∗(c1∗**u1** + c2∗**u2** + c3∗**u3**).
Now the first quadrant and SECT(**v0,u1V1**,**u2V1**,**u3V1**) are related as
v1 − v0 = **v2**•**UV1**
= v2∗(c1∗**u1** + c2∗**u2** + c3∗**u3**)
•(**u1**∗**u1V1** + **u2**∗**u2V1** + **u3**∗**u3V1**)
= v2∗(c1∗**u1V1** + c2∗**u2V1** + c3∗**u3V1**)
so that
viV1 = v2∗ci, i=1,2,3
vV1 = sqrt(**vV1**•**vV1**)
= sqrt(**v2**•**UV1**•**T**[**UV1**]•**v2**)
= v2∗sqrt($c1^2$
+c1∗c2∗**u1V1**•**u2V1**
+c1∗c3∗**u1V1**•**u3V1**
+c2∗c1∗**u2V1**•**u1V1**
+$c2^2$
+c2∗c3∗**u2V1**•**u3V1**
+c3∗c1∗**u3V1**•**u1V1**
+c3∗c2∗**u3V1**•**u2V1**
+$c3^2$)
uV1 = **v1**/vV1
= (v1V1∗**u1V1** + v2V1∗**u2V1** + v3V1∗**u3V1**)/vV1.
as before."

Einstein: "We already know for two vectors **v11** ad **v12** in SECT(**v0,u1V1**,**u2V1**,**u3V1**).
v11 + **v12** = 2∗**v0**
+ (v11V1+ v21V1)∗**u1V1**
+ (v12V1+ v22V1)∗**u2V1**
+ (v13V1+ v23V1)∗**u3V1**
= 2∗**v0** + v1V1∗**u11V1** + v2V1∗**u12V1**
How does this work out in terms of the given origin?"

Breton: "Let
v11 ≡ v11∗(c111∗**u1** + c112∗**u2** + c113∗**u3**)
v12 ≡ v12∗(c121∗**u1** + c122∗**u2** + c123∗**u3**)
where the c's are directional cosines.
Then
v11 + **v12** = (v11∗c111+ v12∗c121)∗**u1**
+ (v11∗c111+ v12∗c122)∗**u2**
+ (v11∗c113+ v12∗c123)∗**u3**
where now, just as vij differs from before, so do the scalars vij∗cijk."

Einstein: "Subtraction, I suspect, differs."

Breton: "Let's see.

$$\mathbf{v11} = \mathbf{v0} + v11V1*\mathbf{u1V1} + v12V1*\mathbf{u2V1} + v13V1*\mathbf{u3V1}$$

$$\mathbf{v12} = \mathbf{v0} + v21V1*\mathbf{u1V1} + v22V1*\mathbf{u2V1} + v23V1*\mathbf{u3V1}$$

so

$$\mathbf{v11} - \mathbf{v12} = (v11V1 - v21V1)*\mathbf{u1V1} + (v12V1 - v22V1)*\mathbf{u2V1} + (v13V1 - v23V1)*\mathbf{u3V1}.$$

Einstein: "This is the same result as subtraction in SECT(**0,u1V1**,**u2V1**,**u3V1**)."

Breton: "I'm curious to see if the same holds for the origin's orientation.

$$\mathbf{v21} = (\mathbf{v11} - \mathbf{v0}) \bullet \mathbf{UV1}^{-1} = v211*\mathbf{u1} + v212*\mathbf{u2} + v213*\mathbf{u3} = v21*\mathbf{uv21} = v21*(c211*\mathbf{u1} + c212*\mathbf{u2} + c213*\mathbf{u3}).$$

$$\mathbf{v22} = (\mathbf{v12} - \mathbf{v0}) \bullet \mathbf{UV1}^{-1} = v221*\mathbf{u1} + v222*\mathbf{u2} + v223*\mathbf{u3} = v22*\mathbf{uv22}$$

So

$$\mathbf{v21} - \mathbf{v22} = (\mathbf{v11} - \mathbf{v0}) \bullet \mathbf{UV1}^{-1} - (\mathbf{v12} - \mathbf{v0}) \bullet \mathbf{UV1}^{-1} = (\mathbf{v11} - (\mathbf{v12}) \bullet \mathbf{UV1}^{-1}$$

just as for SECT(**0,u1V1**,**u2V1**,**u3V1**)."

Einstein: "So while addition is greatly affected by the translation vector, subtraction is unaffected. Breton, draw us an illustration."

With that Breton produced the following illustration.

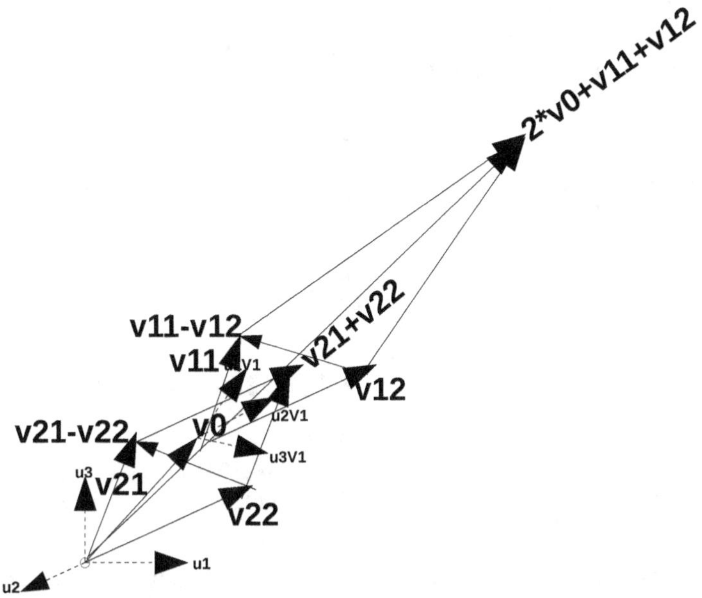

Breton: "The diagram is somewhat crowded. The orientations of the origin are shown in dashed lines with the directions indicated in smaller type. The offset **v0** is shown in the central congested region. The two vectors to be added, **v11** and **v12** are shown in SECT(**v0,u1V1**,**u2V1**,**u3V1**). The same vectors translated into SECT(**0,u1V1**,**u2V1**,**u3V1**) are shown as **v21** and **v22**. The vectors **v11 − v12** and **v21 − v22** are seen to be parallel vectors of equal magnitude. Not shown are the vectors **v0 + v11** and **v0 + v12**. Their sum, however appears as different in both direction and magnitude from **v21 + v22**."

Einstein, intrigued: "Your clarifications should be helpful for the multiplications."

Breton: "Let's see."

$$
\begin{aligned}
\mathbf{v11} * \mathbf{v12} = (&\mathbf{v0} + \mathrm{v11V1}*\mathbf{u1V1} \\
&+ \mathrm{v12V1}*\mathbf{u2V1} \\
&+ \mathrm{v13V1}*\mathbf{u3V1}) \\
&*(\mathbf{v0} + \mathrm{v21V1}*\mathbf{u1V1} \\
&\quad + \mathrm{v22V1}*\mathbf{u2V1} \\
&\quad + \mathrm{v23V1}*\mathbf{u3V1}) \\
= (&\mathbf{v0} + \mathrm{v11V1}*\mathbf{u1V1} \\
&+ \mathrm{v12V1}*\mathbf{u2V1} \\
&+ \mathrm{v13V1}*\mathbf{u3V1})*\mathbf{v0} \\
+ (&\mathbf{v0} + \mathrm{v11V1}*\mathbf{u1V1} \\
&+ \mathrm{v12V1}*\mathbf{u2V1} \\
&+ \mathrm{v13V1}*\mathbf{u3V1})*\mathrm{v21V1}*\mathbf{u1V1}
\end{aligned}
$$

+ (**v0** + v11V1***u1V1**
+ v12V1***u2V1**
+ v13V1***u3V1**)*v22V1***u2V1**
+ (**v0** + v11V1***u1V1**
+ v12V1***u2V1**
+ v13V1***u3V1**)*v23V1***u3V1**).

Einstein: "Enough. The offset surely complicates the result greatly. In SECT(**0,u1V1**,**u2V1**,**u3V1**) the result becomes
v11 * v12 = (v11V1***u1V1**
+ v12V1***u2V1**
+ v13V1***u3V1**)*v21V1***u1V1**
+ (v11V1***u1V1**
+ v12V1***u2V1**
+ v13V1***u3V1**)*v22V1***u2V1**
+ (v11V1***u1V1**
+ v12V1***u2V1**
+ v13V1***u3V1**)*v23V1***u3V1**)
and the result can also be expressed in terms of the origin's orientation in both formulations."

Breton: "Can we agree to consider sections as usually referred to the origin unless we specify otherwise?. Our results will be somewhat limited, while the path to displaced sectors is acknowledged as known but difficult."

Newton, who had followed the discussion critically: "Wouldn't it be simpler to change the origin to **v0** with the same orientation?"

Breton: "What an astute idea. Then all our results for SECT(**0,u1V1**,**u2V1**,**u3V1**) would apply directly.

Einstein: "Then we would have to translate to the original orientation. What good would that do?"

Breton: "It would be like observing from afar as opposed to observing close-up. The same changes observed up close would be observed differently from afar."

Newton: "That sounds like an idea relevant to our purpose."

Breton: "Shall we then agree to consider sections as if they were translated to the origin, or vice-versa for our further discussion."

With relief both Newton and Einstein readily agree.

Einstein: "Let's return to the relationships between the gradients. First give us a definition of sectional gradients."

Breton: "Here it is. You will notice that the gradients can be defined not only for those referenced to, but also for those displaced from the origin."

Definition (basic sectional gradient, divergence, and curl)
Given

$$\mathrm{SECT}(\mathbf{v0},\mathbf{u1V1},\mathbf{u2V1},\mathbf{u3V1}) \text{ at } \mathbf{v1};$$
$$\mathbf{f(v)} \text{ defined in } \mathrm{SECT}(\mathbf{v0},\mathbf{u1V1},\mathbf{u2V1},\mathbf{u3V1});$$
$$\mathbf{dV1} = \mathrm{d1V1}\bullet\mathbf{u1V1} + \mathrm{d2V1}\bullet\mathbf{u2V1} + \mathrm{d3V1}\bullet\mathbf{u3V1},\ \mathrm{diV1}>0;$$

then the **basic sectional gradient** of a generalized function is defined as

$$\mathbf{D}[\mathbf{v0}]\bullet(\mathbf{f(v)};\mathbf{dV1})$$
$$\begin{aligned} \equiv \lim\ (& (\mathbf{f}(\mathbf{v0}+\mathrm{d1V1}\bullet\mathbf{u1V1})-\mathbf{f}(\mathbf{v0}))\bullet\mathbf{u1V1}/\mathrm{d1V1} \\ & + (\mathbf{f}(\mathbf{v0}+\mathrm{d2V1}\bullet\mathbf{u2V1})-\mathbf{f}(\mathbf{v0}))\bullet\mathbf{u2V1}/\mathrm{d2V1} \\ & + (\mathbf{f}(\mathbf{v0}+\mathrm{d3V1}\bullet\mathbf{u3V1})-\mathbf{f}(\mathbf{v0}))\bullet\mathbf{u3V1}/\mathrm{d3V1}) \\ & \qquad \text{as } \mathbf{dV1+} \rightarrow \mathbf{0} \end{aligned}$$

and the related sectional divergence and curl are defined as:

$$\mathrm{D}[\mathbf{v0}]\cdot(\mathbf{f(v)};\mathbf{dV1})$$
$$\begin{aligned} \equiv \lim\ (& (\mathbf{f}(\mathbf{v0}+\mathrm{d1V1}\bullet\mathbf{u1V1})-\mathbf{f}(\mathbf{v0}))\cdot\mathbf{u1V1}/\mathrm{d1V1} \\ & + (\mathbf{f}(\mathbf{v0}+\mathrm{d2V1}\bullet\mathbf{u2V1})-\mathbf{f}(\mathbf{v0}))\cdot\mathbf{u2V1}/\mathrm{d2V1} \\ & + (\mathbf{f}(\mathbf{v0}+\mathrm{d3V1}\bullet\mathbf{u3V1})-\mathbf{f}(\mathbf{v0}))\cdot\mathbf{u3V1}/\mathrm{d3V1}) \\ & \qquad \text{as } \mathbf{dV1+} \rightarrow \mathbf{0} \end{aligned}$$

$$\mathbf{D}[\mathbf{v0}]\wedge(\mathbf{f(v)};\mathbf{dV1})$$
$$\begin{aligned} \equiv \lim\ (& (\mathbf{f}(\mathbf{v0}+\mathrm{d1V1}\bullet\mathbf{u1V1})-\mathbf{f}(\mathbf{v0}))\wedge\mathbf{u1V1}/\mathrm{d1V1} \\ & + (\mathbf{f}(\mathbf{v0}+\mathrm{d2V1}\bullet\mathbf{u2V1})-\mathbf{f}(\mathbf{v0}))\wedge\mathbf{u2V1}/\mathrm{d2V1} \\ & + (\mathbf{f}(\mathbf{v0}+\mathrm{d3V1}\bullet\mathbf{u3V1})-\mathbf{f}(\mathbf{v0}))\wedge\mathbf{u3V1}/\mathrm{d3V1}) \\ & \qquad \text{as } \mathbf{dV1+} \rightarrow \mathbf{0} \end{aligned}$$

end of definition

Newton: "You put the gradient first. Is the divergence related to the gradient as in the first quadrant case?"

Breton: "Yes."

Einstein: "You didn't prove that before. Prove it now."

Breton: "If you wish. I assert

$$\mathrm{D}[\mathbf{v0}]\cdot(\mathbf{f(v)};\mathbf{dV1}) = \mathrm{tr}[\mathbf{D}[\mathbf{v0}]\bullet(\mathbf{f(v)};\mathbf{dV1})]$$

where tr is the trace matrix operator.
Starting with definition

$$\begin{aligned} & (\mathbf{f}(\mathbf{v0}+\mathrm{d1V1}\bullet\mathbf{u1V1})-\mathbf{f}(\mathbf{v0}))\bullet\mathbf{u1V1}/\mathrm{d1V1} \\ & + (\mathbf{f}(\mathbf{v0}+\mathrm{d2V1}\bullet\mathbf{u2V1})-\mathbf{f}(\mathbf{v0}))\bullet\mathbf{u2V1}/\mathrm{d2V1} \\ & + (\mathbf{f}(\mathbf{v0}+\mathrm{d3V1}\bullet\mathbf{u3V1})-\mathbf{f}(\mathbf{v0}))\bullet\mathbf{u3V1}/\mathrm{d3V1} \end{aligned}$$

= (f(v0+d1V1∗u1V1)–f(v0))•I•u1V1/d1V1
+ (f(v0+d2V1∗u2V1)–f(v0))•I•u2V1/d2V1
+ (f(v0+d3V1∗u3V1)–f(v0))•I•u3V1/d3V1
= (f(v0+d1V1∗u1V1)–f(v0))
•[u1∗u1+u2∗u2+u3∗u3]•u1V1/d1V1
+ (f(v0+d2V1∗u2V1)–f(v0))
•[u1∗u1+u2∗u2+u3∗u3]•u2V1/d2V1
+ (f(v0+d3V1∗u3V1)–f(v0))
•[u1∗u1+u2∗u2+u3∗u3]•u3V1/d3V1
= u1•(f(v0+d1V1∗u1V1)–f(v0))∗u1V1•u1/d1V1
+u2•(f(v0+d1V1∗u1V1)–f(v0))∗u1V1•u1/d1V1
+u3•(f(v0+d1V1∗u1V1)–f(v0))∗u1V1•u1/d1V1
+u1•(f(v0+d2V1∗u2V1)–f(v0))∗u2V1•u2/d2V1
+u2•(f(v0+d2V1∗u2V1)–f(v0))∗u2V1•u2/d2V1
+u3•(f(v0+d2V1∗u2V1)–f(v0))∗u2V1•u2/d2V1
+u1•(f(v0+d3V1∗u3V1)–f(v0))∗u3V1•u3/d3V1
+u2•(f(v0+d3V1∗u3V1)–f(v0))∗u3V1•u3/d3V1
+u3•(f(v0+d3V1∗u3V1)–f(v0))∗u3V1•u3/d3V1

For
[tr[D[v0]∗(f(v);dV1)]
= tr[lim u1•(f(v0+d1V1∗u1V1)–f(v0))∗u1V1•u1/d1V1
+u2•(f(v0+d1V1∗u1V1)–f(v0))∗u1V1•u1/d1V1
+u3•(f(v0+d1V1∗u1V1)–f(v0))∗u1V1•u1/d1V1
+u1•(f(v0+d2V1∗u2V1)–f(v0))∗u2V1•u2/d2V1
+u2•(f(v0+d2V1∗u2V1)–f(v0))∗u2V1•u2/d2V1
+u3•(f(v0+d2V1∗u2V1)–f(v0))∗u2V1•u2/d2V1
+u1•(f(v0+d3V1∗u3V1)–f(v0))∗u3V1•u3/d3V1
+u2•(f(v0+d3V1∗u3V1)–f(v0))∗u3V1•u3/d3V1
+u3•(f(v0+d3V1∗u3V1)–f(v0))∗u3V1•u3/d3V1
= lim u1•(f(v0+d1V1∗u1V1)–f(v0))∗u1V1•u1/d1V1
+u2•(f(v0+d2V1∗u2V1)–f(v0))∗u2V1•u2/d2V1
+u3•(f(v0+d3V1∗u3V1)–f(v0))∗u3V1•u3/d3V1]
→D[v0]•(f(v);dV1).

Do you want a proof that
D[v0]∧(f(v);dV1) = c[D[v0]∗(f(v);dV1)]
where **c** is the curl matrix operator, or do you trust me?"

Newton: "We trust you. The definitions imply that the given function has limits *of* the section. The existence of sectional gradients does not necessarily imply either that sectional vector functions have basic directional derivative or even that they are differentiable."

Breton: "True enough. Conversely, the directional derivative
D[v0](f;dV1) ⟵ (f(v0+dV1) – f(v0))/dV1 as dV1 ⟶ 0
may exist in the **dV1** direction even though the sectional gradient does not."

Newton: "So we need a further expansion of the idea of differentiability."

Breton: "Indeed we do. Let me offer this."

Definition (continuous vectorial sectional differentiability)

Given

$$\mathbf{dV1} \equiv \text{dV1}\bullet\mathbf{uV1}$$

$$\equiv \text{d1V1}\bullet\mathbf{u1V1} + \text{d2V1}\bullet\mathbf{u2V1} + \text{d3V1}\bullet\mathbf{u3V1},\ \text{diV1} \geq 0;$$

$\mathbf{f(v)}$, a function over SECT($\mathbf{v0}$,$\mathbf{u1V1}$,$\mathbf{u2V1}$,$\mathbf{u3V1}$)

if for every $\mathbf{uV1}$

$$\mathbf{f(v0+dV1)} - \mathbf{f(v0)} = \mathbf{f(v0+}\text{d1V1}\bullet\mathbf{u1V1)} - \mathbf{f(v0)}$$
$$+ \mathbf{f(v0+}\text{d2V1}\bullet\mathbf{u2V1)} - \mathbf{f(v0)}$$
$$+ \mathbf{f(v0+}\text{d3V1}\bullet\mathbf{u3V1)} - \mathbf{f(v01)}$$
$$\text{as } \mathbf{dV1} \rightarrow \mathbf{0}$$

then

f(v) is **differentiable vectorially** at **v0** in SECT(**v0**,**u1V1**,**u2V1**,**u3V1**).

end of definition

The gradient of a function vectorially differentiable in SECT(**v0**,**u1V1**,**u2V1**,**u3V1**) is called a continuous gradient in SECT(**v0**,**u1V1**,**u2V1**,**u3V1**)."

Einstein, after some scribbling of his own: "This definition works for translated sections!"

Breton: "Yes. They only involve subtractions. Notice that the set of differentiable functions differs from the set of functions with basic gradients which in turn differs from the set of functions with a directional gradient."

Einstein: "A diagram might help here.

So Breton quickly produced the following inclusive diagram.

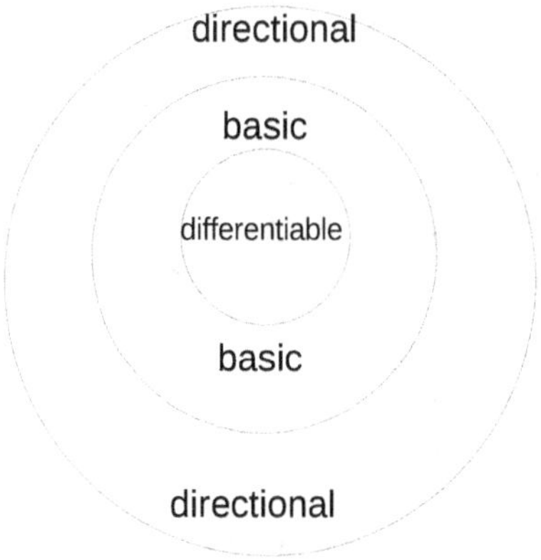

Vector Functions

Newton: "Then for differentiable vector functions we can write

$$\mathbf{f(v0+dV1)} = \mathbf{f(v0)} + \mathbf{dV1} \cdot \mathbf{T[D[}\mathrm{v0}\mathbf{]\bullet(f(v);dV1)}].$$

Breton: "Not generally. Whereas

$$\mathbf{I} = \mathbf{u1 \bullet u1} + \mathbf{u2 \bullet u2} + \mathbf{u3 \bullet u3},$$
$$\mathbf{I} \neq \mathbf{u1V1 \bullet u1V1} + \mathbf{u2V1 \bullet u2V1} + \mathbf{u3V1 \bullet u3V1}$$

generally."

Einstein, looking to lead the discussion: "There must be some relationship between orthogonal gradients and sectional ones."

Newton, insisting: "Please let's focus on why

$$\mathbf{f(v0+dV1)} \neq \mathbf{f(v0)} + \mathbf{dV1} \cdot \mathbf{T[D[}\mathrm{v0}\mathbf{]\bullet(f(v);dV1)}].$$

Breton, selecting the latter path first: "Generally, no such relationship exists. However, for functions which are locally differentiable, we can find one. Our old friend, the scalar triple product, can be useful here. Consider

$$\begin{aligned}
&\mathbf{f(v0+}\mathrm{d}\mathbf{V1)} - \mathbf{f(v0)} \\
&\quad = \mathbf{f(v0+}\mathrm{d1V1}\mathbf{\bullet u1V1)} - \mathbf{f(v0)} \\
&\qquad + \mathbf{f(v0+}\mathrm{d2V1}\mathbf{\bullet u2V1)} - \mathbf{f(v0)} \\
&\qquad + \mathbf{f(v0+}\mathrm{d3V1}\mathbf{\bullet u3V1)} - \mathbf{f(v0)}
\end{aligned}$$

Since

$$\begin{aligned}
&(\mathbf{dV1} \cdot (\mathbf{u2V1 \wedge u3V1})/\mathrm{det}(\mathbf{UV1}) \\
&\qquad = (\mathrm{d1V1}\mathbf{\bullet u1V1})+\mathrm{d2V1}\mathbf{\bullet u2V1})+\mathrm{d3V1}\mathbf{\bullet u3V1}) \\
&\qquad\qquad \cdot (\mathbf{u2V1 \wedge u3V1})/\mathrm{det}(\mathbf{UV1}) \\
&\qquad = (\mathrm{d1V1}\mathbf{\bullet u1V1}) \cdot (\mathbf{u2V1 \wedge u3V1})/\mathrm{det}(\mathbf{UV1}) \\
&\qquad = (\mathrm{d1V1}\mathbf{\bullet}\mathrm{det}(\mathbf{UV1})/\mathrm{det}(\mathbf{UV1}) \\
&\qquad = \mathrm{d1V1}
\end{aligned}$$

and similarly

$$\begin{aligned}
&(\mathbf{dV1} \cdot (\mathbf{u3V1 \wedge u1V1})/\mathrm{det}(\mathbf{UV1}) \\
&\qquad = \mathrm{d2V1} \\
&(\mathbf{dV1} \cdot (\mathbf{u1V1 \wedge u2V1})/\mathrm{det}(\mathbf{UV1}) \\
&\qquad = \mathrm{d3V1},
\end{aligned}$$

$$\begin{aligned}
&\mathbf{f(v0+}\mathrm{d}\mathbf{V1)} - \mathbf{f(v0)} \\
&\quad = (\mathbf{dV1} \cdot (\mathbf{u2V1 \wedge u3V1}) \bullet \mathbf{f(v0+}\mathrm{d1V1}\mathbf{\bullet 1V1)} - \mathbf{f(v0)}/\mathrm{d1V1} \\
&\qquad + \mathbf{dV1} \cdot (\mathbf{u3V1 \wedge u1V1}) \bullet \mathbf{f(v0+}\mathrm{d2V1}\mathbf{\bullet u2V1)} - \mathbf{f(v0)}/\mathrm{d2V1} \\
&\qquad + \mathbf{dV1} \cdot (\mathbf{u1V1 \wedge u2V1}) \bullet \mathbf{f(v0+}\mathrm{d3V1}\mathbf{\bullet u3V1)} - \mathbf{f(v0)}/\mathrm{d3V1})) \\
&\qquad\qquad /\mathrm{det}(\mathbf{UV1})
\end{aligned}$$

Since this looks promising, do it again.

$$\begin{aligned}
&\mathbf{f(v0+}\mathrm{d1V1}\mathbf{\bullet u1V1)} - \mathbf{f(v0)} \\
&\qquad + \mathbf{f(v0+}\mathrm{d2V1}\mathbf{\bullet u2V1)} - \mathbf{f(v0)} \\
&\qquad + \mathbf{f(v0+}\mathrm{d3V1}\mathbf{\bullet u3V1)} - \mathbf{f(v0)} \\
&\quad = (\mathbf{dV1} \cdot (\mathbf{u2V1 \wedge u3V1}) \bullet \mathrm{det}(\mathbf{UV1}) \\
&\qquad\qquad \bullet \mathbf{f(v0+}\mathrm{d1V1}\mathbf{\bullet u1V1)} - \mathbf{f(v0)}/\mathrm{d1V1}
\end{aligned}$$

+dV1 • (u3V1∧u1V1) ∗ det(UV1)
∗ f(v2+d2V1∗u2V1) – f(v0)/d2V1
+dV1 • (u1V1∧u2V1) ∗ det(UV1)
∗ f(v3+d3V1∗u3V1) – f(v0)/d3V1))
/det²[UV1]
= dV1 • ((u2V1∧u3V1) ∗ (u2V1∧u3V1) • u1V1
∗(f(v0+d1V1∗u1V1) – f(v0))/d1V1
+(u3V1∧u1V1) ∗ (u3V1∧u1V1) • u2V1
∗ (f(v2+d2V1∗u2V1) – f(v0))/d2V1
+(u1V1∧u2V1) ∗ (u1V1∧u2V1) • u3V1
∗ (f(v3+d3V1∗u3V1) – f(v0))/d3V1)
/det²[UV1]
= dV1 • ((u2V1∧u3V1) ∗ (u2V1∧u3V1)
• u1V1 ∗(f(v0+d1V1∗u1V1) – f(v0))/d1V1
+(u3V1∧u1V1) ∗ (u3V1∧u1V1)
• u2V1∗ (f(v2+d2V1∗u2V1) – f(v0))/d2V1
+(u1V1∧u2V1) ∗ (u1V1∧u2V1)
• u3V1∗ (f(v3+d3V1∗u3V1) – f(v0))/d3V1)
/det²[UV1]
= dV1 • ((u2V1∧u3V1) ∗ (u2V1∧u3V1)
• (u1V1∗(f(v0+d1V1∗u1V1) – f(v0))/d1V1
+ u2V1∗ (f(v2+d2V1∗u2V1) – f(v0))/d2V1
+ u3V1∗ (f(v3+d3V1∗u3V1) – f(v0))/d3V1)
+(u3V1∧u1V1) ∗ (u3V1∧u1V1)
• (u1V1∗(f(v0+d1V1∗u1V1) – f(v0))/d1V1
+ u2V1∗ (f(v2+d2V1∗u2V1) – f(v0))/d2V1
+ u3V1∗ (f(v3+d3V1∗u3V1) – f(v0))/d3V1)
+(u1V1∧u2V1) ∗ (u1V1∧u2V1)
• (u1V1∗(f(v0+d1V1∗u1V1) – f(v0))/d1V1
+ u2V1∗ (f(v2+d2V1∗u2V1) – f(v0))/d2V1
+ u3V1∗ (f(v3+d3V1∗u3V1) – f(v0))/d3V1)
/det²[UV1]

Now

(u1V1∗(f(v0+d1V1∗u1V1) – f(v0))/d1V1
+ u2V1∗ (f(v2+d2V1∗u2V1) – f(v0))/d2V1
+ u3V1∗ (f(v3+d3V1∗u3V1) – f(v0))/d3V1)
→T[D[v0]∗(f(v);dV1)]

so we finally obtain

f(v0+dV1) – f(v0)
= dV1 • ((u2V1∧u3V1) ∗ (u2V1∧u3V1)
+(u3V1∧u1V1) ∗ (u3V1∧u1V1)
+(u1V1∧u2V1) ∗ (u1V1∧u2V1)
• T[D[v0]∗(f(v);dV1)]
/det²[UV1].

Newton, astonished: "Again not something we could easily guess at."

Einstein, interjecting: "Suppose the section is the first quadrant."

Breton: "Then

$$\begin{aligned}
&\mathbf{u1V1} = \mathbf{u1}\\
&\mathbf{u2V1} = \mathbf{u2}\\
&\mathbf{u3V1} = \mathbf{u3}\\
&\mathrm{d1V1} = \mathrm{dv1}\\
&\mathrm{d2V1} = \mathrm{dv2}\\
&\mathrm{d3V1} = \mathrm{dv3}\\
&\mathbf{dV1} = \mathbf{dv}\\
&\mathbf{u2V1} \wedge \mathbf{u3V1} = \mathbf{u1}\\
&\mathbf{u3V1} \wedge \mathbf{u1V1} = \mathbf{u2}\\
&\mathbf{u1V1} \wedge \mathbf{u2V1} = \mathbf{u3}\\
&\det[\mathbf{UV1}] = 1
\end{aligned}$$

so the relationship becomes

$$\mathbf{f(v0+dv)} - \mathbf{f(v0)} = \mathbf{dv} \cdot (\mathbf{u1} * \mathbf{u1} + \mathbf{u2} * \mathbf{u2} + \mathbf{u3} * \mathbf{u3}) \cdot \mathbf{T[D[v0]*(f(v);dv)]}/1$$

so that

$$\mathbf{f(v0+dv)} = \mathbf{f(v0)} + \mathbf{dv} \cdot \mathbf{T[D[v0]*(f(v);dv)]}$$

which is just what we concluded for a differentiable function over the first quadrant."

Newton: "Little did we realize the usefulness of the scalar triple product."

Relationships between Continuous, Sectional, and Directional Gradients

Einstein, returning to his suggestion: "These gradients all concern a given function. There must be some relationships between the gradients."

Breton: "Indeed there is. Let's start with the transformation

$$\mathbf{UV1} \equiv \mathbf{u1} * \mathbf{u1V1} + \mathbf{u2} * \mathbf{u2V1} + \mathbf{u3} * \mathbf{u3V1}.$$

Since each of the sectional directions may be represented as a vector of directional cosines,"

$$\mathbf{uiV1} = \mathrm{ci1} * \mathbf{u1} + \mathrm{ci2} * \mathbf{u2} + \mathrm{ci3} * \mathbf{u3}, \quad i=1,2,3$$

the transformation may also be represented as

$$[\mathbf{UV1}] = \begin{bmatrix} c11 & c12 & c13 \\ c21 & c22 & c23 \\ c33 & c32 & c33 \end{bmatrix}$$

a matrix of directional cosines."

Newton, encouragingly: "We know

$$\mathbf{UV1}^{-1} = ((\mathbf{u2V1} \wedge \mathbf{u3V1}) \bullet \mathbf{u1} + (\mathbf{u3V1} \wedge \mathbf{u1V1}) \bullet \mathbf{u2} + (\mathbf{u1V1} \wedge \mathbf{u2V1}) \bullet \mathbf{u3}) / \mathrm{det}[\mathbf{UV1}].$$

Einstein, contributing: "For a function with a basic gradient is there a relationship similar to

$$\mathbf{f(v1} + \mathrm{dq} * \mathbf{uv)} \approx \mathbf{f(v1)} + \mathrm{dq} * \mathrm{D}[\mathbf{v1}, \mathbf{v1} + \mathrm{dq} * \mathbf{uv}](\mathbf{f(v)} \,|\, \mathbf{uv}; \mathrm{dv}))$$

which we saw for directional derivatives?"

Breton, taking up the question: "We have seen

$$\begin{aligned} &\mathbf{f(v0+dV1)} - \mathbf{f(v0)} \\ &= \mathbf{dV1} \cdot ((\mathbf{u2V1} \wedge \mathbf{u3V1}) \bullet (\mathbf{u2V1} \wedge \mathbf{u3V1}) \\ &\quad + (\mathbf{u3V1} \wedge \mathbf{u1V1}) \bullet (\mathbf{u3V1} \wedge \mathbf{u1V1}) \\ &\quad + (\mathbf{u1V1} \wedge \mathbf{u2V1}) \bullet (\mathbf{u1V1} \wedge \mathbf{u2V1}) \\ &\quad \cdot \mathbf{T[D}[\mathbf{v0}] \bullet (\mathbf{f(v)}; \mathbf{dV1})] \\ &\quad / \mathrm{det}^2[\mathbf{UV1}] \end{aligned}$$

So let us let $\mathbf{dV1} = \mathrm{diV1} \bullet \mathbf{uiV1}$.
Then

$$\begin{aligned} &\mathbf{f(v0} + \mathrm{diV1} \bullet \mathbf{uiV1)} - \mathbf{f(v0)} \\ &= \mathrm{diV1} \bullet \mathbf{uiV1} \cdot ((\mathbf{u2V1} \wedge \mathbf{u3V1}) \bullet (\mathbf{u2V1} \wedge \mathbf{u3V1}) \\ &\quad + (\mathbf{u3V1} \wedge \mathbf{u1V1}) \bullet (\mathbf{u3V1} \wedge \mathbf{u1V1}) \\ &\quad + (\mathbf{u1V1} \wedge \mathbf{u2V1}) \bullet (\mathbf{u1V1} \wedge \mathbf{u2V1}) \\ &\quad \cdot \mathbf{T[D}[\mathbf{v0}] \bullet (\mathbf{f(v)}; \mathbf{dV1})] \\ &\quad / \mathrm{det}^2[\mathbf{UV1}]. \end{aligned}$$

Newton, protesting: "A somewhat cumbersome result."

Breton: "We can make it simpler. Remember

$$\mathbf{UV1}^{-1} = ((\mathbf{u2V1} \wedge \mathbf{u3V1}) \bullet \mathbf{u1} + (\mathbf{u3V1} \wedge \mathbf{u1V1}) \bullet \mathbf{u2} + (\mathbf{u1V1} \wedge \mathbf{u2V1}) \bullet \mathbf{u3}) / \mathrm{det}[\mathbf{UV1}]$$

so

$$\begin{aligned} \mathbf{UV1}^{-1} \cdot \mathbf{T}[\mathbf{UV1}^{-1}] &= ((\mathbf{u2V1} \wedge \mathbf{u3V1}) \cdot (\mathbf{u2V1} \wedge \mathbf{u3V1}) \\ &\quad + (\mathbf{u3V1} \wedge \mathbf{u1V1}) \cdot (\mathbf{u3V1} \wedge \mathbf{u1V1}) \\ &\quad + (\mathbf{u1V1} \wedge \mathbf{u2V1}) \cdot (\mathbf{u1V1} \wedge \mathbf{u2V1}) \\ &\quad / \mathrm{det}^2[\mathbf{UV1}]. \end{aligned}$$

Then

$$\begin{aligned} &\mathbf{f(v0} + \mathrm{diV1} \bullet \mathbf{uiV1)} - \mathbf{f(v0)} \\ &= \mathrm{diV1} \bullet \mathbf{uiV1} \cdot \mathbf{UV1}^{-1} \cdot \mathbf{T}[\mathbf{UV1}^{-1}] \cdot \mathbf{T[D}[\mathbf{v0}] \bullet (\mathbf{f(v)}; \mathbf{dV1})]. \end{aligned}$$

Newton, favorably: "More memorable!"

Breton: "And it answers Einstein's suggestion as well."

Einstein: "Newton would you make us a table of these various definitions and results."

Newton quickly complied with the request by producing the following table.

Item	Definition/Formula
D[v0]•(f(v);dV1)	lim ((**f(v0+**d1V1•**u1V1**)–**f(v0)**) •**u1V1**/d1V1 + (**f(v0+**d2V1•**u2V1**)–**f(v0)**) •**u2V1**/d2V1 + (**f(v0+**d3V1*****u3V1**)–**f(v0)**) ***u3V1**/d3V1)
T[D[v0]•(f(v);dV1)	lim (**u1V1** •(**f(v0+**d1V1•**u1V1**)–**f(v0)**) /d1V1 + **u2V1** •(**f(v0+**d2V1•**u2V1**)–**f(v0)**) /d2V1 + **u3V1** *(**f(v0+**d3V1*****u3V1**)–**f(v0)**) /d3V1)
UV1	**u1 • u1V1** **+ u2 • u2V1** **+u3 • u3V1**
UV1[-1]	((**u2V1∧u3V1**)•**u1** + (**u3V1∧u1V1**)•**u2** + (**u1V1∧u2V1**)•**u3**) /det[**UV1**]
d1V1	(**dV1** • (**u2V1∧u3V1**) /det(**UV1**)
D[v0]*(f;dV1) • [**UV1**][-1]	((**f(v0+**d1V1*****u1V1**)–**f(v0)**)*****u1** /d1V1 + (**f(v0+**d2V1*****u2V1**) - **f(v0)**)*****u2** /d2V1 + (**f(v0+**d3V1*****u3V1**) -**f(v0)**)*****u3** /d3V1) /det[**UV1**]
f(v0+dV1) – **f(v0)**	(**dV1** • (**u2V1∧u3V1**) • **f(v0+**d1V1•**u1V1**) – **f(v0)**/d1V1 +**dV1** • (**u3V1∧u1V1**) • **f(v2+**d2V1•**u2V1**) – **f(v0)**/d2V1

	+**dV1** • (**u1V1**∧**u2V1**) • **f**(**v3**+d3V1•**u3V1**) – **f**(**v0**)/d3V1)) /det(**UV1**)
f(**v0**+diV1•**uiV1**) – **f**(**v0**)	diV1•**uiV1** • **UV1**$^{-1}$ • **T**[**UV1**$^{-1}$] • **T**[**D**[**v0**]•(**f**(**v**);**dV1**)]

Breton, returning to Einstein's request: "The relationship between a continuous gradient and a sectional gradient is given in the following theorem."

Theorem (relationship between simply continuous and sectional gradients)

Given

f(**v**) a function with a basic, simply continuous gradient at **v0**;

SECT(**v0**,**u1V1**,**u2V1**,**u3V1**) a section at **v0**;

then

D[**v0**]∗(**f**;**dV1**)
= [**D**[**v0**]∗(**f**;**dv**)]
• [**u1V1**∗**u1V1** + **u2V1**∗**u2V1** + **u3V1**∗**u3V1**]
= [**D**[**v0**]∗(**f**;**dv**)] • **T**[**UV1**] • **UV1**.

Proof:

[**D**[**v0**]∗(**f**;**dv**)] • [**u1V1**∗**u1V1** + **u2V1**∗**u2V1** + **u3V1**∗**u3V1**]
= d1V1∗**u1V1** • **T**[**D**[**v0**]∗(**f**;**dv**)]∗**u1V1**/d1V1
+ d2V1∗**u2V1** • **T**[**D**[**v0**]∗(**f**;**dv**)]∗**u2V1**/d2V1
+ d3V1∗**u3V1** • **T**[**D**[**v0**]∗(**f**;**dv**)]∗**u3V1**/d3V1
≈ (**f**(**v0**+d1V1∗**u1V1**) – **f**(**v0**))∗**u1V1**/d1V1
+ (**f**(**v0**+d2V1∗**u2V1**) – **f**(**v0**))∗**u2V1**/d2V1
+ (**f**(**v0**+d3V1∗**u3V1**) – **f**(**v0**))∗**u3V1**/d3V1

for continuously spatially differentiable function with basic gradients
⟶ **D**[**v0**]∗(**f**;**dV1**).

Moreover,

T[**UV1**] • **UV1** = [**u1V1**∗**u1V1** + **u2V1**∗**u2V1** + **u3V1**∗**u3V1**]

qed

Breton: "The inverse relationship
D[**v0**]∗(**f**;**dv**)
= **D**[**v0**]∗(**f**;**dV1**)
• [**u1V1**∗**u1V1**+ **u2V1**∗**u2V1**+ **u3V1**∗**u3V1**]$^{-1}$
also holds for functions with basic, simply continuous gradients at **v0**. On the other hand, sectional gradients may also be related to directional derivatives."

Einstein: "Show us."

Breton complied with the request by writing down the following theorem with its proof.

Theorem (relationship between sectional gradients and directional derivatives)

Given

$$\mathrm{SECT}(\mathbf{v0,u1V1,u2V1,u3V1}),\ \text{a section at } \mathbf{v0};$$
$$\mathrm{CQ}\ \text{a curve into } \mathrm{SECT}(\mathbf{v0,u1V1,u2V1,u3V1})\ \text{at } \mathbf{v0} \equiv \mathbf{v}(\mathrm{q1});$$
$$\mathbf{f(v)}\ \text{a function sectionally differentiable in } \mathrm{SECT}(\mathbf{v0,u1V1,u2V1,u3V1})\ \text{at } \mathbf{v0};$$
$$\mathbf{dV1} \equiv \mathrm{dV1} \ast \mathbf{udV1}$$
$$\equiv \mathrm{d1V1} \ast \mathbf{u1V1} + \mathrm{d2V1} \ast \mathbf{u2V1} + \mathrm{d3V1} \ast \mathbf{u3V1};$$
$$= \mathbf{v}(\mathrm{q1+dq}) - \mathbf{v}(\mathrm{q1});$$

then

$$\mathrm{D[q1,q1+dq]}(\mathbf{f}\,|\,\mathrm{CQ;dq})$$
$$= \mathrm{D[q1,q1+dq]}(\mathbf{v}(\mathrm{q})\,|\,\mathrm{CQ;dq}) \bullet [\mathbf{UV1}^{-1}] \bullet \mathbf{T}[\mathbf{UV1}^{-1}] \bullet \mathbf{T}[\mathbf{D}[\mathbf{v0}] \ast (\mathbf{f};\mathbf{dV1})].$$

Proof:

$$\mathrm{D[q1,q1+dq]}(\mathbf{v}(\mathrm{q});\mathrm{dq})$$
$$\approx (\mathbf{v}(\mathrm{q1+dq}) - \mathbf{v}(\mathrm{q1}))/\mathrm{dq}$$
$$= \mathrm{dV1} \ast \mathbf{udV1}/\mathrm{dq}$$
$$= (\mathrm{d1V1} \ast \mathbf{u1V1} + \mathrm{d2V1} \ast \mathbf{u2V1} + \mathrm{d3V1} \ast \mathbf{u3V1})/\mathrm{dq}$$

Then,

$$\mathrm{D[q1,q1+dq]}(\mathbf{v}(\mathrm{q});\mathrm{dq}) \bullet [\mathbf{UV1}^{-1}] \bullet \mathbf{T}[\mathbf{UV1}^{-1}] \bullet \mathbf{T}[\mathbf{D}[\mathbf{v0}] \ast (\mathbf{f};\mathbf{dV1})]$$
$$\approx (\mathrm{d1V1} \ast \mathbf{u1V1} + \mathrm{d2V1} \ast \mathbf{u2V1} + \mathrm{d3V1} \ast \mathbf{u3V1})$$
$$\bullet [(\mathbf{u2V1} \wedge \mathbf{u3V1}) \ast \mathbf{u1} + (\mathbf{u3V1} \wedge \mathbf{u1V1}) \ast \mathbf{u2} + (\mathbf{u1V1} \wedge \mathbf{u2V1}) \ast \mathbf{u3}]$$
$$\bullet [\mathbf{u1} \ast (\mathbf{u2V1} \wedge \mathbf{u3V1}) + \mathbf{u2} \ast (\mathbf{u3V1} \wedge \mathbf{u1V1}) + \mathbf{u3} \ast (\mathbf{u1V1} \wedge \mathbf{u2V1})]$$
$$\bullet [\mathbf{u1V1} \ast (\mathbf{f}(\mathbf{v0} + \mathbf{d1V1} \ast \mathbf{u1V1}) - \mathbf{f}(\mathbf{v0}))/\mathrm{d1V1}$$
$$+ \mathbf{u2V1} \ast (\mathbf{f}(\mathbf{v0} + \mathbf{d2V1} \ast \mathbf{u2V1}) - \mathbf{f}(\mathbf{v0}))/\mathrm{d2V1}$$
$$+ \mathbf{u3V1} \ast (\mathbf{f}(\mathbf{v0} + \mathbf{d3V1} \ast \mathbf{u3V1}) - \mathbf{f}(\mathbf{v0}))/\mathrm{d3V1}]$$
$$/((\det[\mathbf{UV1}])^{2} \ast \mathrm{dq})$$
$$= (\mathbf{f}(\mathbf{v0} + \mathrm{d1V1} \ast \mathbf{u1V1}) - \mathbf{f}(\mathbf{v0})$$
$$+ \mathbf{f}(\mathbf{v0} + \mathrm{d2V1} \ast \mathbf{u2V1}) - \mathbf{f}(\mathbf{v0})$$
$$+ \mathbf{f}(\mathbf{v0} + \mathrm{d3V1} \ast \mathbf{u3V1}) - \mathbf{f}(\mathbf{v0}))/\mathrm{dq}$$
$$\approx (\mathbf{f}(\mathbf{v0} + \mathbf{dV1}(\mathbf{q1})) - \mathbf{f}(\mathbf{v0}))/\mathrm{dq}$$
$$\longrightarrow \mathrm{D[q1,q1+dq]}(\mathbf{f}\,|\,\mathbf{v}(\mathrm{q});\mathrm{dq}).$$

qed

Breton: "Sectional gradients of vector functions are transformations. As transformations they may be used meaningfully only restrictedly."

Newton: "Explain what you mean."

Breton: "As transformations

$$\mathbf{UV1}: \ V3 \rightarrow V3$$
$$\mathbf{UV1}^{-1}: V3 \rightarrow V3$$

In restriction

$$\mathbf{UV1}: \quad \text{positive quadrant} \rightarrow \text{SECT}(\mathbf{0,u1V1},\mathbf{u2V1},\mathbf{u3V1})$$
$$\mathbf{UV1}^{-1}: \text{SECT}(\mathbf{0,u1V1},\mathbf{u2V1},\mathbf{u3V1}) \rightarrow \text{positive quadrant}$$

The use of these transformations restricted to the latter domains is called their **canonical usage**. Only canonical usage is appropriate for sectional gradients. For other than canonical usage, the transformations will generate non-canonical images and so be inappropriate for sectional gradients."

Einstein, leading by questioning: "How does this affect functions?"

Breton: "Let F be the image of the gradient of **f(v)**. Then for a function differentiable in SECT(**v0**,**u1V1**,**u2V1**,**u3V1**)

$$[\mathbf{UV1}]^{-1} \bullet \mathbf{T}[\mathbf{UV1}]^{-1} \bullet \mathbf{T}[\mathbf{D}[\mathbf{v0}]{*}(\mathbf{f};\mathbf{dV1})]:$$
$$\text{SECT}(\mathbf{v0},\mathbf{u1V1},\mathbf{u2V1},\mathbf{u3V1}) \rightarrow F$$

so that

$$\mathbf{T}[\mathbf{UV1}]^{-1} \bullet \mathbf{T}[\mathbf{D}[\mathbf{v0}]{*}(\mathbf{f};\mathbf{dV1})]: \text{positive quadrant} \rightarrow F.$$

Einstein: "It appears that sectional gradients must be extensions of first quadrant gradients. Is this so?"

Breton: "If **u1V1** = **u1**, **u2V1** = **u2**, and **u3V1** = **u3**, then **UV1=I** the identity transformation; SECT(**v0,u1V1**,**u2V1**,**u3V1**) is the positive quadrant; and **D[v0]***(**f**;**dV1**) = **D[v0]***(**f**;**dv**) for the positive quadrant. Differences in the eight quadrants are thus selected by the sign of **uiV1**. When all eight quadrants have identical sectional gradients, the gradient is then **simply continuous at v0**."

Einstein: "We might as well have translated the origin to **v0**, since any vector can serve as an origin. Then the section and the origin would have a common reference."

Newton, parrying by opening another path: "On the other hand a section can be so narrow it looks like a direction. When **uiV1**→**uv**, i=1,2,3

$$\mathbf{UV1} = \mathbf{u1}{*}\mathbf{uv} + \mathbf{u2}{*}\mathbf{uv} + \mathbf{u3}{*}\mathbf{uv}$$
$$= (\mathbf{u1}+\mathbf{u2}+\mathbf{u3}){*}\mathbf{uv},$$

an outer product. Then **UV1** has no inverse."

Breton, concluding: "Then the sectional gradient becomes

$$\mathbf{D}[\mathbf{v1}]{*}(\mathbf{f};\mathbf{dV1}) = \lim\, (\mathbf{f}(\mathbf{v1}+dv1{*}\mathbf{uv})-\mathbf{f}(\mathbf{v1})){*}\mathbf{uv}/dv1$$
$$+ (\mathbf{f}(\mathbf{v1}+dv2{*}\mathbf{uv})-\mathbf{f}(\mathbf{v1})){*}\mathbf{uv}/dv2$$
$$+ (\mathbf{f}(\mathbf{v1}+dv3{*}\mathbf{uv})-\mathbf{f}(\mathbf{v1})){*}\mathbf{uv}/dv3$$
$$= 3{*}((\mathbf{f}(\mathbf{v1}+dv{*}\mathbf{uv})-\mathbf{f}(\mathbf{v1}))/dv){*}\mathbf{uv}$$
$$\rightarrow 3{*}\mathbf{D}[\mathbf{v1},\mathbf{v1}+dv{*}\mathbf{uv}]{*}(\mathbf{f}(\mathbf{v});dv{*}\mathbf{uv}).$$

or given **v**(q), a non–stalled process curve with

$$\mathbf{v}(q1) = \mathbf{v1}$$

and $$\mathbf{v}(q1+dq) = \mathbf{v1}+dv*\mathbf{uv}.$$

D[v1]*(f;dV1)

$$= 3*D[\mathbf{v1},\mathbf{v1}+\mathbf{dv}(\mathbf{v1},\mathbf{v}(x1+dx))](\mathbf{f}(\mathbf{v});dv(\mathbf{v1},\mathbf{v}(x)))*\mathbf{uv}$$

$$= 3*D[q1,q1+dq](\mathbf{f}|\mathbf{v}(q);dq) / D[q1,q1+dq](dv(\mathbf{v1},\mathbf{v}(q);dq)) *\mathbf{uv}.$$

Sectional gradients are then reduced to outer products."

Einstein, also concluding: "Sectional gradients thus occupy an intermediate position between continuous, unrestricted gradients and directional gradients."

Newton, looking again to direct the conversation: "These gradients are also transformations. What is their rank?"

Breton: "Let's talk about the rank of transformations directly."

Rank of a Transformation

Breton: "As an idea, **V3** is so defined that vectors, considered no matter how large or how small, or in whatever direction, are never lacking. This property arises from the properties of Q, the field of quotient numbers underlying **V3**. This sufficiency underlies the conclusion that the positive quadrant gradient **D[v0]*(v;dv)** = **I**, the identity matrix.

It is possible, however, to designate specific subsets S of **V3** over which the gradient does not equate to the identity matrix. Restricted to surfaces, curves or points, **D[v0]*(v**|S;**dv)** is singular, that is det[**D[v0]*(v**|S;**dv)**] = 0. For example, for constant **v0**,

$$\mathbf{D[v0]*(v0;dv)} = [\mathbf{0}]$$

for which det[**D[v0]*(v0;dv)**] = 0.

The restriction may also be written as a function **f**:**V3**→**S** where the restriction is seen as the image of **f**. The function itself may have a basic gradient restricted to the positive quadrant, or a section, or even just a direction."

With that Breton created the following table of the various possibilities corresponding to the rank of the gradient as a matrix:

Rank	Determinant	Restriction/Image
0	0	points
1	0	curves
2	0	surfaces
3	non–zero	vicinities

Rank of Matrices in V3

Einstein, exploring: "Give us a few examples"

Breton: "If $\mathbf{f}=\mathbf{c}$, $\mathbf{c}$ constant, then

$$\mathbf{D}[v]*(\mathbf{c};\mathbf{dv})] = \mathbf{0}.$$
$$\det[\mathbf{0}] = 0$$

This is a rank 0 transformation.

If $\mathbf{f}=f*\mathbf{uc}$, $\mathbf{uc}$, a fixed direction, then

$$\mathbf{D}[v,v+dv\bullet\mathbf{uc}]*(f*\mathbf{uc};\mathbf{dv}) = (\mathbf{D}[v,v+dv\bullet\mathbf{uc}]*(f;\mathbf{dv}))*\mathbf{uc}$$
$$\det[(\mathbf{D}[v,v+dv\bullet\mathbf{uc}]*(f;\mathbf{dv}))*\mathbf{uc}]= \mathbf{0}$$

This is a rank 1 gradient.

If $\mathbf{f}=f1*\mathbf{uc1}+f2*\mathbf{uc2}$, two fixed directions, then

$$\mathbf{D}[\mathbf{v0}]*(\mathbf{f1}*\mathbf{uc1}+\mathbf{f2}*\mathbf{uc2};\mathbf{dv}) = (\mathbf{D}[v,v+dv\bullet\mathbf{uc1}]*(f1;\mathbf{dv}))*\mathbf{uc1} + (\mathbf{D}[v,v+dv\bullet\mathbf{uc}]*(f2;\mathbf{dv}))*\mathbf{uc2}$$
$$\det[\mathbf{D}[\mathbf{v0}]*((\mathbf{D}[v,v+dv\bullet\mathbf{uc1}]*(f1;\mathbf{dv}))*\mathbf{uc1} + (\mathbf{D}[v,v+dv\bullet\mathbf{uc}]*(f2;\mathbf{dv}))*\mathbf{uc2}] = \mathbf{0}$$

This is a rank 2 gradient..

If $\mathbf{f}=f1*\mathbf{u1}+f2*\mathbf{u2}+f3*\mathbf{u3}$, then $\det[\mathbf{D}[\mathbf{v0}]*(\mathbf{f};\mathbf{dv})],\neq,\mathbf{0}$, rank 3."

Einstein: "You give us more than examples; they are explanations."

Breton: "We see that derivatives in **V3** follow a pattern. Newton would you list our results generically."

Local Differentiation of Sums and Products

Newton: "The results below hold within a class for continuous, positive quadrant, sectional, or directional local derivatives (given appropriate differentiability). A formula cast specifically for continuous gradients yields specific results for continuous gradients, not directional or sectional gradients, etc.

Item:	D• (divergence)	D∧ (curl)	D∗ (gradient)
c constant			0
c constant	0	0	[0]
f+g, sum			**D∗**f + **D∗**g
f+g, sum	D•**f** + D•**g**	**D**∧f + **D**∧g	**D**∗f + **D**∗g
f∗g, product			f∗**D∗**g + g∗**D∗**f
1/f, reciprocal			−(**D**∗f)/f²
f∗**g**, product	f∗D•**g** + **g**•**D∗**f	f∗**D∧g** +**g∧(D∗**f)	f∗**D∗g** + **g∗D∗**f
f•g, product			**f•D∗g** + **g•D∗f**
f∧g, product	**f•D∧g − g•D∧f**	**f•**T[**D∗g**] − **g•**T[**D∗f**] − **f∗D•g** + **g∗D•f**	**C(f)•D∗g** − **C(g)•D∗f**
f∗g, product	**f∗D•g** +**g•T(D∗f)**		

Sums and Products of non-process Derivatives

Breton: "The table can be used to determine other combinations. For instance for

A = u1∗a1+u2∗a2+u3∗a3

D•[A] = (D•**a1**)∗**u1** + (D•**a2**)∗**u2** + (D•**a3**)∗**u3**
D•[T[A]] = u1•T[D∗a1] + u2•T[D∗a2] + u3•T[D∗a3]
D•(f•A) = (f•u1)∗D•**a1 + (f•u2)∗**D•**a2 + (f•u3)∗**D•**a3**
+ a1•(u1•D∗f) + a2•(u2•D∗f) + a3•(u3•D∗f)
D•(f•T[A]) = u1•(f•D∗a1 + a1•D∗f)
+ u2•(f•D∗a2 + a2•D∗f)
+ u3•(f•D∗a3 + a3•D∗f)
D∧(f•A) = (f•u1)∗D∧a1 + (f•u2)∗D∧a2 + (f•u3)∗D∧a3
+ a1∧(u1•D∗f) + a2∧(u2•D∗f) + a3∧(u3•D∗f)
D∧(f•T[A]) = u1∧(f•D∗a1) + u2∧(f•D∗a2) + u3∧(f•D∗a3)
+ (u1∧a1 + u2∧a2 + u3∧a3)•D∗f
D∗(f•A) = (f•u1)∗D∗a1 + (f•u2)∗D∗a2 + (f•u3)∗D∗a3
+ (a1∗u1 + a2∗u2 + a3∗u3)•D∗f
D∗(f•T[A]) = u1∗(f•D∗a1) + u2∗(f•D∗a2) + u3∗(f•D∗a3)
+ (u1∗a1 + u2∗a2 + u3∗a3)•D∗f.

Newton: "Note that for **C** the curl matrix operator

D•(C(f)) = D∧f
D•T(C(f)) = −D∧f
D∧(D∧f) = D•T[D∗f] − D•D∗f
D∧(f∧g) = D•T(f∗g) − D•(f∗g).

Breton: "Combinations of these local derivatives are also possible. For instance, there often appear

$$\mathrm{D}\bullet(\mathbf{D}\wedge\mathbf{f}) = 0$$

$$\mathbf{D}\wedge(\mathbf{D}*\mathbf{f}) = \mathbf{0}.$$

The combination $\mathrm{D}\bullet\mathbf{D}*$ is likewise often useful."

Einstein, seizing the initiative: "We have investigated many derivatives of the set of vectors. What have you to say about integrals?"

Breton: "Each of the derivatives is matched by an integral. We can appreciate this by following a parallel path up our mountain."

Local Integration

Breton: "In **V3** integration, like differentiation, may be defined with reference either to the underlying field Q or to the set of vectors itself.

Integration by means of scalar increments produces three possible integrals:

in a given direction and so called **directional**
along v(q) **with respect to** q
along v(q) **with respect to v**.

Integration by means of vector increments also arises from three possibilities:

V3→Q called the **invergence**, symbolized by **I**•
V3→**V3** called the **incurl**, symbolized by **I**∧
and Q→**V3** called the **ingradient**, symbolized by **I**∗

Integration in **V3** arises from the topology of **V3** restricted or unrestricted."

Newton: "Where do we start?"

Breton: "The parallel path up the mountain suggests starting with directional integrals, just as we did with derivatives."

Einstein: "Then start by giving us a definition."

Directional Integrals

Breton: "Of course. Here it is.

Definition (directional integrals)

Given
v1, a vector in **V3**;
v2, another vector in **V3**;
uv ≡ **u**(**v2**−**v1**), a direction;
f(**v**) a given vector function;
dv a positive quotient number;
then if it exists

lim dv∗(**f**(**v1**) + **f**(**v1**+dv∗**uv**) + **f**(**v1**+2∗dv∗**uv**) +...+ **f**(**v2**))
as dv→0

is called the
directional integral of f in the uv direction from v1 to v2.
end of definition

Directional integrals are symbolized as
I[**v1**,**v2**](**f**(**v**) | **uv**;dv).

Newton: "How is your definition a limit? It looks like a summation of more and more values."

Breton: "That's correct. But the finite sum is multiplied by dv, a multiplicative factor which decreases as the sum increases. So a meaningful limit may result from such a process."

Einstein: "Illustrate what you mean."

Breton: "Certainly. Consider a constant vector function, $\mathbf{f(v)} = \mathbf{f0}$.
Starting with dv=1 and $\mathbf{v2=2*v1}$

$$1*(\mathbf{f(v1)} + \mathbf{f(v1}+dv*\mathbf{uv})) = \mathbf{f0 + f(v1+v1)} = 2*\mathbf{f0}$$

$$(1/2)*(\mathbf{f(v1)} + \mathbf{f(v1}+(1/2)*\mathbf{uv}) + \mathbf{f(v2)}) = 3*\mathbf{f0}/2$$

$$(1/3)*(\mathbf{f(v1)} + \mathbf{f(v1}+(1/3)*\mathbf{uv}) + \mathbf{f(v1}+(2/3)*\mathbf{uv}) + \mathbf{f(v2)}) = 4*\mathbf{f0}/3$$

$$\ldots$$

$$(1/n)*(\mathbf{f(v1)} + \mathbf{f(v1}+(1/n)*\mathbf{uv}) + \mathbf{f(v1}+(2/n)*\mathbf{uv})$$
$$+\ldots+ \mathbf{f(v2)}) = (n+1)*\mathbf{f0}/n$$

You can see that the limiting value of this sequence of values is $1*\mathbf{f0}$. So even though the value of the sums increases, it is balanced by a value which decreases, so that the ratio of the two may stabilize to a limit."

Newton: "This is just the inverse of the derivative where the decreasing numerator is multiplied by 1/dv so that the ratio of the two may stabilize to a limit."

Breton, reflecting: "We see here an instance of something seen often not only in mathematics but also in nature. Often enough things are in balance. We see it here where balanced ratios come to a definite limit."

Einstein: "And sometimes not."

Newton: "Resulting then in chaos."

Breton: "There is something divine about balance. We see it imperfectly in our lives, even while contemplating a perfect prospect. Here is a little poem I've composed for you my friends."

Balance

O Holy Trinity
Father, Son, and Spirit
In balance perfect

pleased to look on Mary
singing sweet lullabies
to infant Jesus

pleased with Jesus the man
obedient in death
hammered to the Cross

commanding lovingly
dead Jesus to arise
glorious from death

inspiring Jesus' Church,
His Mystical Body,
throughout place and time.

Newton: "How is it that such diversity nevertheless expresses such unity?"

Einstein, refocusing: "Still there are differences. The definitions for derivatives worked only for basic functions. Integrals are not bound by such restrictions."

Breton: "Very little get by you Einstein. You are correct as the perusal of the definitions clearly shows. Integrals may be defined over functions which are more general than basic. However, for functions with basic derivatives special results may be achieved."

Einstein: "Show us."

Breton: "For functions with a basic directional derivative in direction **uv** from **v1** to **v2**

$$\begin{aligned}
&\mathrm{I}[\mathbf{v1},\mathbf{v2}](\mathrm{D}[\mathbf{v1},\mathbf{v1}+\mathbf{dv}](\mathbf{f}(\mathbf{v})|\mathbf{uv};dv)|\mathbf{uv};dv)\\
&\quad = \lim dv*(\mathrm{D}[\mathbf{v1},\mathbf{v1}+\mathbf{dv}](\mathbf{f}(\mathbf{v})|\mathbf{uv};dv)\\
&\qquad\qquad + \mathrm{D}[\mathbf{v1}+\mathbf{dv},\mathbf{v1}+2*\mathbf{dv}](\mathbf{f}(\mathbf{v})|\mathbf{uv};dv)\\
&\qquad\qquad + \ldots\\
&\qquad\qquad + \mathrm{D}[\mathbf{v2},\mathbf{v2}+\mathbf{dv}](\mathbf{f}(\mathbf{v})|\mathbf{uv};dv))\\
&\quad = \lim dv*((\mathbf{f}(\mathbf{v1}+\mathbf{dv})|\mathbf{uv}-\mathbf{f}(\mathbf{v1})|\mathbf{uv})/dv\\
&\qquad\qquad + (\mathbf{f}(\mathbf{v1}+2*\mathbf{dv})|\mathbf{uv}-\mathbf{f}(\mathbf{v1}+\mathbf{dv})|\mathbf{uv})/dv\\
&\qquad\qquad + \ldots\\
&\qquad\qquad + (\mathbf{f}(\mathbf{v2}+dv*\mathbf{uv})|\mathbf{uv}-\mathbf{f}(\mathbf{v2})|\mathbf{uv})/dv)\\
&\quad = \lim (\mathbf{f}(\mathbf{v2}+dv*\mathbf{uv})|\mathbf{uv}-\mathbf{f}(\mathbf{v1})|\mathbf{uv})
\end{aligned}$$

Consequently for functions with basic directional derivatives in direction **uv** from **v1** to **v2**,

$$\begin{aligned}
&\lim \mathbf{f}(\mathbf{v2}+dv*\mathbf{uv})|\mathbf{uv}\\
&\quad = \mathbf{f}(\mathbf{v1}) + \mathrm{I}[\mathbf{v1},\mathbf{v2}](\mathrm{D}[\mathbf{v1},\mathbf{v1}+\mathbf{dv}](\mathbf{f}(\mathbf{v})|\mathbf{uv};dv)|\mathbf{uv};dv).
\end{aligned}$$

Newton: "That is a beautiful analogy to the proposition

$$\mathbf{f}(\mathbf{v1}+\mathbf{dv}(\mathbf{v1})) \approx \mathbf{f}(\mathbf{v1}) + dq*\mathrm{D}[\mathbf{v1},\mathbf{v1}+\mathbf{dv}](\mathbf{f}(\mathbf{v})|CV;dq).$$

Breton: "Do you remember the fundamental theorem for integral calculus?"

Newton: "Of course."

Breton: "The above equation is the **fundamental theorem of integral calculus for directional integrals**."

Newton: "Yes, I see that. We have climbed much higher up the mountain, and the view is spectacular."

Einstein, spoiling the view: "Surely there are other integrals than directional integrals."

Process Integration

Breton: "Yes, the better view shows us the way more clearly. The integral of a restricted function **f**|CQ is defined from the induced topology. Such integrals are called **integrals along CQ with respect to** q."

Definition (integrals along CQ with respect to q)

Given

CQ = {**v**}, a curve;

f|CQ, a vector function restricted to CQ;

then if it exists

lim dq∗(**f**(q1) + **f**(q1+dq) + **f**(q1+2∗dq) + ... + **f**(q2))

as dq→0

is called the

integral of f|CQ **from f**(q1) **to f**(q2).

end of definition:

Integrals of restricted functions along CQ with respect to q are symbolized as

I[q1,q2](**f**(q)|CQ;dq)).

For restricted functions with a basic derivative along CQ from q1 to q2

I[q1,q2](D[q,q+dq](**f**(q)|CQ;dq)|CQ;dq)
= lim dq∗(D[q1,q1+dq](**f**(q)|CQ;dq)
+ D[q1+dq,q1+2∗dq](**f**(q)|CQ;dq)
+ ...
+ D[q2,q2+dq](**f**(q)|CQ;dq))
= lim dq∗((**f**(q1+dq)−**f**(q1))/dq
+ (**f**(q1+2∗dq)−**f**(q1+dq))/dq
+ ...
+ (**f**(q2+dq)−**f**(q2))/dq)
= lim (**f**(q2+dq)−**f**(q1)).

Consequently for functions restricted to CQ with basic directional derivatives along the curve from q1) to q2),

lim **f**(q2+dq)
= **f**(q1) + **I**[q1,q2](D[q,q+dq](**f**(q)|CQ;dq)|CQ;dq).

This equation is the **fundamental theorem of integral calculus for integrals of restricted functions over v**(q) **with basic directional derivatives** along the curve from **v**(q1) to **v**(q2)."

If the curve branches, integrals may be assigned to each separate branch."

Einstein, trying to dampen his admiration: "How about stalled processes?"

Breton: "For a process stalled from **f**(q1) to **f**(q2),

D[**f**(q) | CQ;dq) = **0**

f(q2) = **f**(q1)

I[q1,q2](D[q,q+dq](**f**(q) | CQ;dq) = **0**

even though **I**[q1,q2](**f**(q) | CQ;dq) may not be zero.

Newton, imitating Einstein: "Why not process integrals with reference only to the set of vectors."

Breton: "You know the way; it is well marked.

Definition (integrals along CV with respect to **v**)

Given

CV, a curve in V3;

v1 a location in CV;

ut(**v**), the curve direction at **v** in CV;

for **v** in CV define

dv(**v**) ≡ dv•**ut**(**v**), dv >0;

f(**v**) | CV, a vector function over CV;

then if it exists

lim dv∗(**f**(**v1**) + **f**(**v1**+**dv**(**v1**))

+ **f**(**v1**+**dv**(**v1**)+ **dv**(**v1**+**dv**(**v1**)))

+ ...

+ **f**(**v**2)) as dv→0

is called the

integral of f(v) from v1 to v2 along CV.

end of definition

Integrals of compound functions along CV are symbolized as

I[**v1**,**v2**](**f**(**v**) | CV;dv).

The end-point of the integration may then be computed as

v2 = **v1** + **I**[**v1**,**v2**](**dv**(**v**) | CV;dv).

For functions with basic derivatives along CV from **v1** to **v2**

I[**v1**,**v2**](D[**v**,**v**+**dv**(**v**)](**f**(**v**) | CV;dv) | CV;dv)

= lim dv∗(D[**v1**,**v1**+**dv**(**v1**)](**f**(**v**) | CV;dv)

+ D[**v1**+**dv**(**v1**), **v1**+**dv**(**v1**)+**dv**(**v1**+**dv**(**v1**))](**f**(**v**) | CV;dv)

+ ...

+ D[**v2**,**v2**+**dv**(**v2**)](**f**(**v**) | CV;dv))

= lim **f**(**v1**+**dv**(**v1**)) | CV–**f**(**v1**)

+ **f**(**v1**+**dv**(**v1**)+**dv**(**v1**+**dv**(**v1**))) | CV–**f**(**v1**+**dv**(**v1**)) | CV

+ ...

+ **f**(**v2**+**dv**) | CV–**f**(**v2**)

= lim **f**(**v2**+**dv**(**v2**)) | CV–**f**(**v1**).

Consequently for compound functions with basic derivative of **f(v)** along CV from **v1** to **v2**,
lim **f(v2+dv(v2))**|CV
= **f(v1)** + **I[v1,v2]**(D[**v,v+dv(v)**](**f(v)**|CV;dv)|CV;dv).
The above equation is the **fundamental theorem of integral calculus along CV with respect to v**."

Newton: "These many integrals all bear a certain similarity."

Breton: "So generic formulas for integrating sums and products of functions, provided they apply to the same class of integrals, can be written as:

I[]((**f1**+**f2**)|u) = **I**[](**f1**|u) + **I**[](**f2**|u)
I[](D[](**f1**•**f2**|u)|u) = I[](**f1**•D[](**f2**|u)|u)
+ I[](D[](**f1**|u)•**f2**|u)
I[](D[](**f1**∧**f2**|u)|u) = **I**[](**f1**∧D[](**f2**|u)|u)
+ **I**[](D[](**f1**|u)∧**f2**|u)
I[](D[](f1∗f2|u)|u) = **I**[](f1∗D[](f2|u)|u)
+ **I**[](D[](f1|u)∗f2|u).

For a constant vector **c**,

I[](**c**•**f**|u) = **c**•**I**[](**f**|u),
I[](**c**∧**f**|u) = **c**∧**I**[](**f**|u)
I[](**c**∗**f**|u) = **c**∗**I**[](**f**|u).

Also,

abs(**I**[](**f**|u)) ≤ I[](abs(**f**)|u)).

Einstein: "I suspect a parallel path exists for invergences, incurls and ingradients."

Breton: "Just so. Here are the definitions.

Definition (directional invergences, incurls, and ingradients)
Given
v1, a vector in V3;
v2, another vector in V3;
uv ≡ **u(v2−v1)**, a direction;
dv ≡ dv∗**uv**;
then
the forward directional invergence, incurl and ingradient in direction **uv** from **v1** to **v2** are defined as:

Invergence:
lim (**f(v1)** + **f(v1+dv)**|**uv** + **f(v1**+2∗**dv)**|**uv** +... + **f(v2)**)•**dv**,
as dv→0

Incurl:
lim (**f(v1)** + **f(v1+dv)**|**uv** + **f(v1**+2∗**dv)**|**uv** +... + **f(v2)**)∧**dv**,
as dv→0

Ingradient:
lim (**f(v1)** + **f(v1+dv)|uv** + **f(v1+2*dv)|uv**+... + **f(v2))*dv**,
as dv→0
or for scalar functions
lim (f(**v1**) + f(**v1+dv**)|**uv** + f(**v1+2*dv**)|**u** +... + f(**v2**))***dv**,
as dv→0.
end of definition

Directional invergences, incurls and ingradients are symbolized as

I[v1,v2]•(f(v)|uv;dv)
I[v1,v2]∧(f(v)|uv;dv)
I[v1,v2]*(f(v)|uv;dv).

Newton: "How is it that you can write the equations so quickly?"

Breton: "From the definitions a kind of similarity appears. So by substituting certain symbols one definition can be transformed into another."

Einstein: "Show us."

Breton: "Here are two tables which can be useful for making the transitions."

final/initial	directional	along CQ	along CV
directional	**v**+dq***uv**	**v**+dq***uv** for **v**(q+dq)	**v**+dq***uv** for **v+dv**
along CQ	**v**(q+dq) for **v**+dq***uv**	**v**(q+dq)	**v**(q+dq) for **v+dv**
along CV	**v+dv** for **v**+dq***uv**	**v+dv** for **v**(q+dq)	**v+dv**

Substitution table for process derivatives

final/initial	directional	along CQ	along CV
directional	n*dv***uv**	n*dv***uv** for **v**(q1+n*dq)	n*dv***uv** for (**vn**+**dv**)
along CQ	**v**(q1+n*dq) for n*dv***uv**	**v**(q1+n*dq	**v**(q1+n*dq for (**vn**+**dv**)
along CV	(**vn**+**dv**) for n*dv***uv**	(**vn**+**dv**) for **v**(q1+n*dq)	(**vn**+**dv**)

Substitution table for process integrals

Einstein: "Surely restrictions to basic functions apply."

Breton: "Of course."

Einstein: "Then proceed to non-process derivatives along curves."

Breton: "We are treading easy ground. Integration aimed at inverting divergences, curls and gradients along curves may also be defined. Here are the definitions."

Invergences, Incurls, and Ingradients along curves

Definition (invergences, incurls and ingradients along CQ)

Given

CQ = {**v**(q)}, a curve;

dq >0;

dv(q) ≡ **v**(q+dq) − **v**(q);

then the directional invergence, incurl and ingradient along CQ from **v**(q1) to **v**(q2) are defined as:

Invergence:

lim (**f**(**v**(q1))•**dv**(q1)
+ **f**(**v**(q1+dq))•**dv**(q1+dq)
+ **f**(**v**(q1+2*dq))•**dv**(q1+2*dq)
+...
+ **f**(**v**(q2)))•**dv**(q2), as dq→0

Incurl:

lim (**f**(**v**(q1))∧**dv**(q1)
+ **f**(**v**(q1+dq))∧**dv**(q1+dq)
+ **f**(**v**(q1+2*dq))∧**dv**(q1+2*dq)
+...
+ **f**(**v**(q2)))∧**dv**(q2), as dq→0

Ingradient:
lim $(\mathbf{f}(\mathbf{v}(q1))*\mathbf{dv}(q1)$
$+ \mathbf{f}(\mathbf{v}(q1+dq))*\mathbf{dv}(q1+dq)$
$+ \mathbf{f}(\mathbf{v}(q1+2*dq))*\mathbf{dv}(q1+2*dq)$
+...
$+ \mathbf{f}(\mathbf{v}(q2)))*\mathbf{dv}(q2)$, as $dq \to 0$
or for scalar functions
lim $(f(\mathbf{v}(q1))*\mathbf{dv}(q1)$
$+ f(\mathbf{v}(q1+dq))*\mathbf{dv}(q1+dq)$
$+ f(\mathbf{v}(q1+2*dq))*\mathbf{dv}(q1+2*dq)$
+...
$+ f(\mathbf{v}(q2)))*\mathbf{dv}(q2)$, as $dq \to 0$.
end of definition

Invergences, incurls and ingradients along CQ are symbolized as

$I[\mathbf{v}(q1),\mathbf{v}(q2)]\bullet(\mathbf{f}(\mathbf{v})\,|\,CQ;\mathbf{dv})$
$\mathbf{I}[\mathbf{v}(q1),\mathbf{v}(q2)]\wedge(\mathbf{f}(\mathbf{v})\,|\,CQ;\mathbf{dv})$
$\mathbf{I}[\mathbf{v}(q1),\mathbf{v}(q2)]*(\mathbf{f}(\mathbf{v})\,|\,CQ;\mathbf{dv})$.

Einstein: "And for curves defined with reference to the vector set only?"

Breton: "We have similar definitions.

Definition (invergences, incurls and ingradients along CV)
Given
CV = a continuous curve;
$dv > 0$;
$\mathbf{dv}(\mathbf{v}) \equiv dv*\mathbf{ut}(\mathbf{v})$, $\mathbf{v}$ in CR;
then
the directional invergence, incurl, and ingradient along CV from **v1** to **v2** are defined as:

Invergence:
lim $\mathbf{f}(\mathbf{v1})\bullet\mathbf{dv}(\mathbf{v1})$
$+ \mathbf{f}(\mathbf{v1}+\mathbf{dv}(\mathbf{v1}))\bullet\mathbf{dv}(\mathbf{v1}+\mathbf{dv}(\mathbf{v1}))$
+...
$+ \mathbf{f}(\mathbf{v2})\bullet\mathbf{dv}(\mathbf{v2})$, as $dv \to 0$

Incurl:
lim $\mathbf{f}(\mathbf{v1})\wedge\mathbf{dv}(\mathbf{v1})$
$+ \mathbf{f}(\mathbf{v1}+\mathbf{dv}(\mathbf{v1}))\wedge\mathbf{dv}(\mathbf{v1}+\mathbf{dv}(\mathbf{v1}))$
+...
$+ \mathbf{f}(\mathbf{v2})\wedge\mathbf{dv}(\mathbf{v2})$, as $dv \to 0$

Ingradient:
lim f(v1)*dv(v1)
+ f(v1+dv(v1))*dv(v1+dv(v1))
+...
+ f(v2)*dv(v2), as dv→0
or for scalar functions
lim f(v1)*dv(v1) + f(v1
+dv(v1))*dv(v1+dv(v1))
+...
+ f(v2)*dv(v2), as dv→0.
end of definition

Invergences, incurls and ingradients along CV are symbolized as
I[v1,v2]•(f(v)|CV;dv)
I[v1,v2]∧(f(v)|CV;dv)
I[v1,v2]*(f(v)|CV;dv).
Since abs(dv(v)) = dv, we may define
I[v1,v2]*(f(v)|CV;abs(dv))
≡ lim f(v1)*abs(dv(v1))
+ f(v1+dv(v1))*abs(dv(v1+dv(v1)))
+...
+ f(v2)*abs(dv(v2)) as dv→0.

Newton: "Substitution is easier than for process definitions. The symbols, •, ∧, * need only to be interchanged.

Einstein: "Do you have something comparable to derivatives for functions multiplied with constants?"

Breton: "Indeed we do. For constant c1, c1 and c2, c2 with f1, f1 and f2, f2 defined on CQ or CV corresponding to the condition u
I[]•((c1*f1+c2*f2)|u;dv)
= c1*I[]•(f1|u;dv) + c2*I[]•(f2|u;dv)
I[]•((c1*f1+c2*f2)|u;dv)
= c1•I[]*(f1|u;dv) + c2•I[]*(f2|u;dv)
I[]•((c1∧f1+c2∧f2)|u;dv)
= c1•I[]∧(f1|u;dv) + c2•I[]∧(f2|u;dv)
I[]•((c1*f1+c2*f2)|u;dv)
= c1*I[]•(f1|u;dv) + c2*I[]•(f2|u;dv)
I[]∧((c1*f1+c2*f2)|u;dv)
= c1*I[]∧(f1|u;dv) + c2*I[]∧(f2|u;dv)
I[]∧((c1*f1+c2*f2)|m1;dv)
= c1∧I[]*(f1|m1;dv) + c2∧I[]*(f2|m1;dv)
I[]∧((c1∧f1+c2∧f2)|u;dv)
= c1•T[I[]*(f1|u;dv) – c1•I[]*(f1|u;dv)
+ c2•T[I[]*(f2|u;dv) – c2•I[]*(f1|u;dv)

I[]*((c1*f1+c2*f2)|u;dv)
= c1*I[]*(f1|u;dv) + c2*I[]*(f2|u;dv)
I[]*((c1•f1+c2•f2)|u;dv)
= c1•I[]*(f1|u;dv) + c2•I[]*(f2|u;dv)
I[]*((c1*f1+c2*f2)|u;dv)
= c1*I[]*(f1|u;dv) + c2*I[]*(f2|u;dv)
I[]*((c1*f1+c2*f2)|u;dv)
= c1*I[]*(f1|u;dv) + c2*I[]*(f2|u;dv)
I[]*((c1∧f1+c2∧f2)|u;dv)
= C(c1)•I[C]*(f1|u;dv) + C(c2)•I[C]*(f2|u;dv)
where C is the curl matrix operator."

Einstein: "So there are remarkable parallels between derivatives and integrals. Are there no interactions between the two?"

Breton: "You ask an interesting question. For a sub-class of restricted functions the following theorem applies."

Theorem (interior cancellation over curves)
Given:
CQ, a curve;
f(v(q)), a function with has a basic gradient along CQ;
dv(qi) ≡ v(qi+dq) – v(qi) for v(qi) in CQ;
then

I[v(q1),v(qm)]•(D[v(q),v(q+dq)]*(f(v)|CQ;dv)|CQ;dv)
= lim f(v(qm)+dv(qm)) – f(v(q1)) as dv→0.

Proof:
I[v(q1),v(qm)]•(D[v(qi)]*(f(v(q))|Cq;dv)|CQ;dv)
= lim I[C(v(q1),v(qm))]•((f(v(qi)+dv(qi))–f(v(qi)))
*qd(dv(qi))|CQ;dv)
= lim (f(v(q1)+dv(q1))–f(v(q1)))*qd(dv(q1))•dv(q1)
+ (f(v(q1+dv(q1)+dv(q1+dq))–f(v(q1+dv(q1)))
*qd(dv(q1+dq))•dv(q1+dq)
+ ...
+ (f(v(vm)+dv(vm))–f(v(vm)))*qd(dv(vm))•dv(qm)
= lim f(v(q1)+dv(q1))–f(v(q1))
+ f(v(q1+dq)+dv(q1+dq))–f(v(q1+dv(q1)))
+ ...
+ f(v(vm)+dv(vm))–f(v(vm))
= lim (f(v(vm)+dv(vm)) – f(v(q1))) as dv→0.

qed

Newton: "So this is interior cancellation where only the end points of the integration count. A similar result must exist for the same curves as functions of CV."

Breton: "Correct. We can express the result as a corollary to the theorem."

Corollary
Given:
CV, a curve;
f(v), a function with have a basic gradient along CV;
dv(v) ≡ dv∗**ut(v)**, **v** in CV;
then

I[**v1**,**vm**]•(**D**[**v**,**v**+**dv(v)**]∗(**f(v)**|CV;**dv**)|CV;**dv**)
= lim (**f(vm)**+**dv(vm)**)) - **f(v1)**, as dv→0.

These equations are the **fundamental theorem of integral vector calculus extended to curves**.

Newton: "What a splendid development."

Einstein, acknowledging but continuing: "When. in tp1.1 we discussed similar issues, Breton introduced step functions. They proved useful then. Have they a place in the context of these local integrals?"

Step functions over curves.

Breton: "Since they are real functions, the unit step functions defined in tp1.1 may be multiplied as scalar functions with vector functions."

Newton: "Remind us of step functions.

Breton: "Remember in tp1.1 we saw that they play a part in the construction of integrals over Q. We might find similar results for the set of vectors. Here is the basic definition.
Consider two functions, defined as

$$u(q-q1) \equiv 0, \quad q \le q1$$
$$\equiv 1, \quad q > q1$$

and

$$v(q-q1) \equiv 0, \quad q < q1$$
$$\equiv 1, \quad q \ge q1.$$

The functions u(q) and v(q) are called **unit** step functions. They have equal and constant values everywhere except at q=q1 which is called their **stepping point**. At the stepping point, they differ both in their values and in their limits.

Einstein: "Show us a diagram."

Breton quickly sketched the following two illustrations which he handed to his friends..

Breton: "Look at this sketch of u(q−q1).

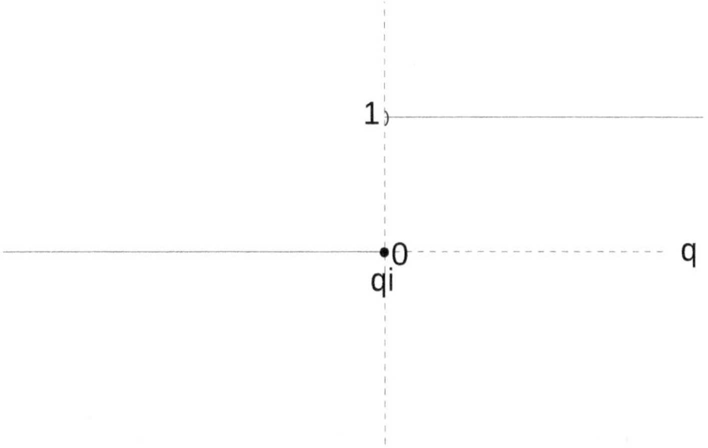

u(q−q1)

The dashed lines show the axes, one for q and the other for the value of the function. The solid lines show the value of the function. At all values of q less than and including zero, u(q−q1) = 0. For q greater than 0, but not including zero u(q−q1) = 1

Now look at the sketch of v(q−q1).

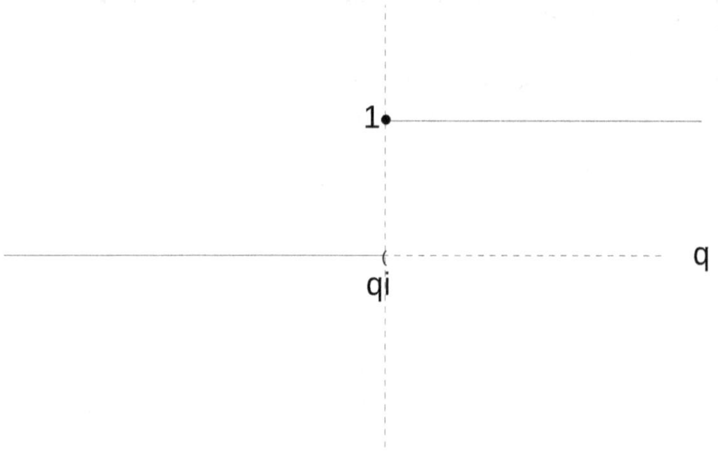

v(q−q1)

The values of v(q−q1) are identical to u(q−q1), except at q=q1. At q1 a reverse situation holds. At all values of q less than but not including zero, v(q−q1) = 0. For q greater than 0, including zero v(q−q1) = 1.

Newton: "Of what use are these functions?

Einstein, joining in objecting: “The v stepping function is too easily confused with a vector **v**.”

Breton: "One is a scalar, the other a vector.”

Newton: "But the magnitude of the vector is also symbolized as a scalar v. Breton, your symbology here is ambiguous. Choose another symbol!”

Breton: "But a similar argument might be made disfavoring the symbol u for a unit step function. It could be easily confused with a unit vector **u**.”

Now a vigorous discussion arose among the friends about a proper symbol. It was finally agreed that no single symbol like h or n would do. So a double character would be employed. The friends finally agreed to the following. symbols for these stepping functions.

$$\mathrm{us}(q-q1) \equiv 0,\ q \le q1$$
$$\equiv 1,\ q > q1$$

and

$$\mathrm{vs}(q-q1) \equiv 0,\ q < q1$$
$$\equiv 1,\ q \ge q1.$$

Breton, taking the lead: "Functions
restricted to a direction **uv** with reference to **v**
restricted to a curve CQ with reference to q
or restricted to a connected curve CV with reference to **v**
preserve some order. The previous definitions for unit step functions can be expanded accordingly."

Definition 1 (step functions along curves)

Given

q, a quotient number;

for

v1, a constant vector;
uv, a given direction;
v = **v1** + q***uv**, a line of vectors;

$$\mathrm{us}(q - qi)|\mathbf{uv} \equiv 1,\ q - qi > 0$$
$$\equiv 0,\ q - qi \le 0$$
$$\mathrm{vs}(q - qi)|\mathbf{uv} \equiv 1,\ q - qi \ge 0$$
$$\equiv 0,\ q - qi < 0$$

for

v(q) a curve over CQ

$$\mathrm{us}(\mathbf{v}(q) - \mathbf{v}(qi))|CQ \equiv 1,\ q - qi > 0$$
$$\equiv 0,\ q - qi \le 0$$
$$\mathrm{vs}(\mathbf{v}(q) - \mathbf{v}(qi))|CQ \equiv 1,\ q - qi \ge 0$$
$$\equiv 0,\ q - qi < 0$$

for **v** in CV, a curve from **v1** to **v2**

$$\mathrm{us}(\mathbf{v} - \mathbf{vi})|CV \equiv 1,\ \mathbf{v} \text{ in } (\mathbf{vi},\mathbf{v2}]$$
$$\equiv 0,\ \mathbf{v} \text{ in } [\mathbf{v1},\mathbf{vi}]$$
$$\mathrm{vs}(\mathbf{v} - \mathbf{vi})|CV \equiv 1,\ \mathbf{v} \text{ in } [\mathbf{vi},\mathbf{v2}]$$
$$\equiv 0,\ \mathbf{v} \text{ in } [\mathbf{v1},\mathbf{vi}).$$

end of definition

Newton, intrigued: "So why multiply these functions with vector functions?"

Step Functions multiplied with vector functions

Breton: "Let me write a few results for you. First consider the result of differentiating the step functions.
Given

the generic condition *c* as **uv** or CQ or CV;
q generically either the directional variable
or the one in CQ or CV;

then

$$D[q](us(t-qi)|c;d_f t) = 0, \quad q \neq qi$$
$$= \lim (1/dq) \quad q = qi$$
$$D[q](vs(t-qi)|c;d_b t) = 0, \quad q \neq qi$$
$$= \lim (1/dq) \quad q = qi;$$

Einstein: "Explain please."

Breton: "If q≠qi, both step functions have a constant value, either 0 or 1. The derivative there is thus zero.
At q=qi

$$D[q](us(t-qi)|c;d_f t) = \lim ((us(qi + dq)-us(qi))/dq)$$
$$= \lim (1 - 0)/dq$$

while

$$D[q](us(t-qi)|c;d_b t) = \lim ((us(qi)-us(qi - dq))/dq)$$
$$= \lim (0 - 0)/dq.$$

In contrast

$$D[q](vs(t-qi)|c;d_b t) = \lim ((vs(qi)-vs(qi - dq))/dq)$$
$$= \lim (1 - 0)/dq.$$
$$D[q](vs(t-qi)|c;d_f t) = \lim ((vs(qi + dq)-vs(qi))/dq)$$
$$= \lim (1 - 1)/dq$$

Newton: "The forward derivative of one is the same as the backward derivative of the other."

Breton: "The non-zero derivative is called an **impulse** function. As you can see the same impulse function can arise from two different step functions."

Newton, musing to himself: "The difference arises because us(qi) = 0 while vs(qi) = 1."

Einstein, reflecting as well: "Which is just opposite of their limits

$$\lim_f (us(qi + dq)-us(qi)) = 1$$

while

$$\lim_f ((vs(qi)-vs(qi - dq)) = 0.$$

Breton, continuing the meditation: "Also,

$$\lim_b ((us(qi)-us(qi - dq)) = 0$$

and

$$\lim_b (vs(qi + dq)-vs(qi)) = 1.$$

These stepping functions provide a strong sense of direction along curves."

Newton, pushing forward: "So how are they used with functions?"

Breton: "For **f** (q)|c, a generalized function, where c represents any of the conditions, directional, or along a curve
(**f**(q)|c)∗D[q](us(t−qi);d_ft)
= **f**(qi)∗D[qi](us(t−qi);d_ft), q=qi
= **0**, otherwise."

Einstein: "Explain please."

Breton: "For q≠qi,
D[qi](us(t−qi);d_ft) = 0, so the product also equals 0.
For q=qi only the value **f**(qi) figures in the product."

Newton, concluding: "So this product selects a single functional value from a vector function."

Einstein, suddenly enlightened: "This should have an impact on integration."

Breton: "Indeed, it does.
I[](**f**(q)•D[q](us(t−qi)|c;d_ft)|c;dq)
= **f**(qi)
I[](**f**(q)•D[q](vs(t−qi)|;d_bt)|c;dq)
= **f**(qi)
The integral of the products reduce in the limit to
I[](**f**(q)•D[q](us(t−qi)|c;d_ft)|c;dq) = **f**(qi)•(1/dq)•dq.

Newton, reflecting: "Just a slight change in notation produces a much different result.
I[](**f**(q)•D[q](us(t−qi)|c;d_bt)|c;dq) = 0
and
I[](**f**(q)•D[q](vs(t−qi)|;d_ft)|c;dq) = 0.

Breton: "A different symbol, however subtle the difference, for a different idea."

Newton, leading by imitating Einstein: "How about another product?"

Breton: "In addition to the above we can have
us(q−qi)∗D[q](**f**(t)|c;d_fq)
= **0** q≤qi
= D[q](**f**(t);d_fq) q>qi
D[q](**f**(t)∗us(t−qi)|c;d_ft)
= **0** q<qi
= lim **f**(qi+dq)|c/dq q=qi
= D[q](**f**(t);d_fq) q>qi.

Newton: "Yes, these products can be easily accepted."

Breton: " Then how about the following integrals?
I[](**f**(q)•D[q](us(t-qi)|c;d_ft)|c;dq)
= **f**(qi)
I[](**f**(q)•D[q](vs(t-qi)|c;d_bt)|c;dq)
= **f**(qi).

Newton: "Yes, these are also acceptable. The integral cancels out the derivative."

Breton: "The derivatives, acting as impulse functions, select only one value of the function for the integral."

Einstein: "How about results for special functions?"

Breton: "Let's start with functions which are
basic
continuous,
bounded,
asymptotic."

Einstein: "We know about basic and continuous, but what about bounded, or asymptotic?"

Breton: "Bounded simply means that the functional values are everywhere finite. Asymptotic means that the functional values collapse to zero as q $\longrightarrow \pm\infty$."

Einstein: "Illustrate please."

Breton: "Here is a simple illustration."

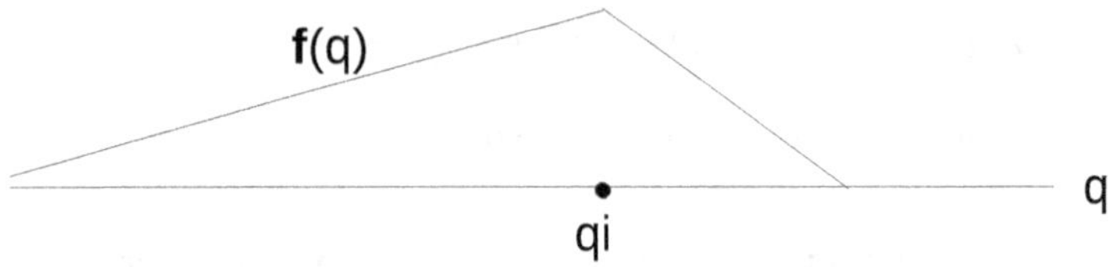

The function falls off to zero in the positive and negative directions (asymptotic); the functional values are always finite (bounded); the function is continuous, though not necessarily its derivatives (as at **f**(qi)."

Einstein: "And so for these functions what results do we have?"

Breton: "We can write
$\mathbf{I}[\,]((us(t-qi))\bullet D[q](\mathbf{f}(t)|c;d_f t)|c;dq)$
$= -\mathbf{f1}$, limit of $\mathbf{f}$ from above as $q \to qi$
$\mathbf{I}[\,]((vs(t-qi))\bullet D[q](\mathbf{f}(t)|c;d_b t)|c;dq)$
$= -\mathbf{f0}$, limit of $\mathbf{f}$ from below as $q \to qi$
$\mathbf{I}[\,](D[q]((us(t-qi))\bullet \mathbf{f}(t)|c;d_f t)|c;dq)$
$= \mathbf{0}$
$\mathbf{I}[\,](D[q]((vs(t-qi))\bullet \mathbf{f}(t)|c;d_b t)|c;dq)$
$= \mathbf{0}$.

Newton: "How can you put down the results so quickly?"

Breton: "Let's take the first equation. The function us(t−qi) limits the integral to values greater than qi. The integral could then be written as

$\mathbf{I}[qi+dq,\infty]D[q](\mathbf{f}(t)|c;d_f t)|c;dq)$

For a basic function, interior cancellation holds, so that
$\mathbf{I}[qi,\infty]D[q](\mathbf{f}(t)|c;d_f t)|c;dq) = \mathbf{f}(\infty) - \mathbf{f}(qi+dq)$.
Since f is asymptotic, $\mathbf{f}(\infty) = \mathbf{0}$. The limit of $\mathbf{f}(qi+dq)$ is $\mathbf{f1}$."

Einstein: "Which is just $\mathbf{f}(qi)$."

Newton: "But this would work for functions which are continuous everywhere except for a jump at qi."

Breton: "Just so."

Einstein: " Illustrate what you mean."

Breton: "Consider this illustration."

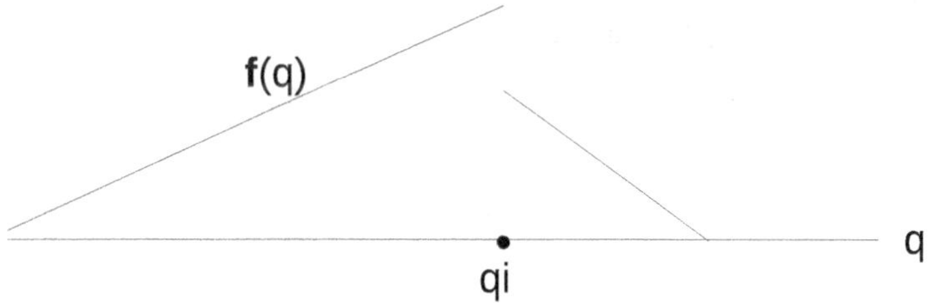

The function jumps down at qi. The integral picks out the limit of the function from above (not necessarily the functional value itself)."

Newton: "And a similar result comes from

$\mathbf{I}[\,]((vs(t-qi))\bullet D[q](\mathbf{f}(t)|c;d_b t)|c;dq)$

which picks out the limit from below."

Einstein, curious: "Let me change the problem just a little. What is
I[](D[q]((us(t-qi))•**f**(t)|c;d_ft)|c;dq)?"

Breton: "Now we need to differentiate a product. From above we have

D[q]((us(t-qi))•**f**(t)|c;d_ft)
= lim(us(q+dq-qi))•**f**(q+dq) - us(q-qi))•**f**(q))/dq
= **0** $q<qi$
= lim **f**(qi+dq)|c/dq $q=qi$
= D[q](**f**(t);d_fq) $q>qi$

so

I[](D[q]((us(t-qi))•**f**(t)|c;d_ft)|c;dq)
= lim lim **f**(qi+dq)|c/dq + D[qi+dq](**f**(t);d_fq)
+ D[i+2•dq](**f**(t);d_fq)
+ D[qi+3•dq](**f**(t);d_fq)
+ ...)*dq
= lim lim **f**(qi+dq)|c/dq + (**f**(qi+2•dq) - **f**(qi+dq))/dq
+ (**f**(qi+3•dq) - **f**(qi+2•dq))/dq
+ ...)*dq
= **0**.

for a basic, asymptotic function."

Newton: "A similar argument would show
I[](D[q]((vs(t-qi))•**f**(t)|c;d_bt)|c;dq)
= **0**.

Breton: "Yes. You see the analogy."

Einstein, inquiringly: "Suppose further **f** is discontinuous at qi?"

Breton: "We know from above
I[]((us(t-qi))•D[q](**f**(t)|c;d_ft)|c;dq) = **-f1.**
What is
I[](**f**(t)*(D[q](us(q-qi)|;d_ft)|c;dq)?"

Einstein: "Now we're dealing with the impulse function so
I[](**f**(q)*D[t](us(t-qi)|;d_ft)|c;dq)
= **f1**
So
I[](**f**(q)*D[t](us(t-qi)|;d_ft)|c;dq)
= -**I**[](D[q](**f**(t)|;d_ft)*us(q-qi)|c;dq).

Newton: "And likewise,
I[](**f**(q)*D[t](vs(t-qi)|;d_bt)|c;dq)
= -**I**[](D[q](**f**(t)|;d_bt)*vs(q-qi)|c;dq).

Einstein: "These step functions do indeed force us to reason closely. Are there other uses for them?"

Breton: "The integral over [+∞,−∞] may be convenient mathematically but hardly qualifies as something amenable to Theoretical Physics. Consider instead functions over a portion of a curve from **v**(q0) to **v**(qm). Further multiply the function with a step function at **v**(q0). Then

I[**v**(q0),**v**(qm)](D[**v**(q),**v**(q+dq)](**f**(**v**(q))•us(**v**(q)-**v**(q0))
|CQ;dq)|CQ;dq)
= lim **f**(**v**(qm+dq))|CQ - **f**(**v**(q0)).

Contrast this with

I[](D[**v**(q),**v**(q+dq)](**f**(**v**(q))•us(**v**(q)-**v**(q0))
|CQ;dq)|CQ;dq)
=- **f**(**v**(q0)).

Newton, joining exuberantly: "It is easy to show how the result for the other step function,

I[**v**(q0),**v**(qm)](D[**v**(q-dq),**v**(q)](f(**v**(q))•vs(**v**(q)-**v**(q0))
|CQ;dq)|CQ;dq)
= **f**(**v**(qm))–lim **f**(v(q0 - dq))|CQ.

Newton, continuing: "Surely **f** must be continuous. If so, then all the functional values cancel except **f**(**v**(qm+dq)) and **f**(**v**(q0)). This is an impressive role for step functions."

Breton: "Much same can be said for the curve defined from the set of vectors only. Let **v1** = **v**(q0) and **vm** = **v**(qm). Then

I[**v1**,**vm**](D[**v**,**v**+**dv**,](**f**(**v**)•us(**v**-**v1**)|CV;dv)|CV;dv)
= lim **f**(**vm**+**dv**(**vm**))|CV -**f** (**v1**)

and

I[**v1**,**vm**](D[**v**-**dv**,**v**](**f**(**v**)•vs(**v**-**vi**)|CV;dv)|CV;dv)
= **f**(**vm**)–lim **f**(**v1**-**dv**(**v1**))|CV.

Newton: "Why use the step functions at all? Why not just use the finite limits of integration without the step function?"

Breton: "Tell me is

I[**v1**,**vm**](D[**v**,**v**+**dv**,](**f**(**v**)•us(**v**-**v1**)|CV;dv)|CV;dv)

the same as

I[**v1**,**vm**](D[**v**,**v**+**dv**,](**f**(**v**)|CV;dv)|CV;dv)?"

Newton: "Yes, both equal **f**(**vm**) -**f**(**v1**)."

Breton: "But suppose = lim **f**(**vm**+**dv**(**vm**))|CV does not equal **f**(**vm**). Isn't this a possibility? In such a case the two integrals would not have the same value."

Newton: "I never thought of that."

Breton: "The step functions enable us to specify the condition at the end points of the integration, whether the functional value or a limiting value is to be included. As you can imagine, this applies both the upper and lower end-points."

Einstein: "Impressive, very impressive."

Breton: "Impressed? You might be even more impressed by the following.
Generally, if

f1(q*uv) is a continuous function in the direction uv;
g(q), the anti-derivative of f1,
that is, D[q](g(t*uv)|uv;dt) = f1(q*uv);
q0<qi<qm;
d constant;

then the function

f(q*uv) = f1(q*uv) + d*us(q-qi)

is integrated as

I[q0*uv,qm*uv]((f(q*uv) + d*us(q-qi))|uv;dq)
= g(qm*uv) - g(q0*uv) + d*(qm-qi).

Further if

f1(v(q)) is a continuous function along CQ;
g(vq)), the anti-derivative of f1;
that is, D[v(q),v(q+dq)](g(v(t))|CQ;dt) = f1(v(q));
q0<qi<qm;
d constant;

then the function

f(v(q)) = f1(v(q)) + d*us(q-qi)|CQ

is integrated as

I[v(q0),v(qm)]((f(v(q))|CQ + d*us(q-qi)|CQ)|CQ;dq)
= g(v(qm)) - g(v(q0)) + d*(qm-qi).

And if

f1(v) is a continuous function along CV;
g(v), the anti-derivative of f1, that is,
D[v+dv](g(v)|CV;dv) = f1(v);
v1 in [v0,vm];
d constant;

then the function

f(v) = f1(v) + d*us(v-vi)|CV

is integrated as

I[q0,qm]((f(v)|CV + d*us(v-qi)|CV)|CV;dq)
= g(vm) - g(v0) + d*I[qi,qm](1|CV;dq).

Einstein: "Illustrate please."

Breton: "Let me illustrate simply by drawing in just two dimensions. You will have to imagine the third."

Breton took out a pad of paper and quickly produced the following sketch.

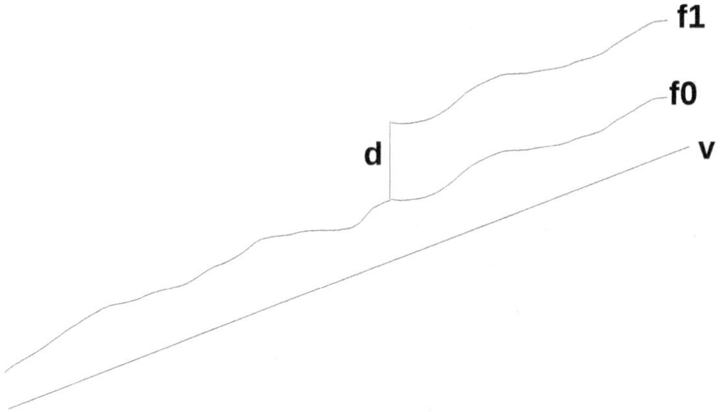

The line marked **v** could be either a directional line, or a curve. The function **f0** is shown as a continuous function over **v**. At some point **vi** (not shown) a step is introduced with the vector **d**. The addition of the step produces a new, discontinuous function, **f1**."

Newton: "In this way the target functions for valid integration are very greatly expanded."

Breton: "Just so. Consequently, we can also reflect that the above integrals are the **fundamental theorem of integral calculus of functions over curves in V3 extended to functions with a finite step discontinuity**."

Einstein: "The unity of our development is impressive indeed."

Newton: "Let me think about this a little. In addition to derivatives, step functions must also possess gradients."

Einstein: "We also know that functions with similar functional values may have different gradients."

Breton: "We need a definition. Here it is."

Process Gradients of Step Functions

Definition (basic process gradients of step functions)

Given

q, a quotient variable;

for

v1, a constant vector;
uv, a given direction;
v = **v1** + q***uv**, a line of vectors;
dv ≡ dq•**uv**;

then

the basic **directional gradient of a directional step function in direction uv** is defined as

D[v,v+dv]•(us(q − qi)|**uv;dv**)
≡ lim(us(q+dq − qi)|**uv** − us(q − qi)|**uv**)•**qd**(**dv**)
= **0**, q ≠ qi
= **uv**/dq, q = qi

D[v,v+dv]•(vs(q − qi)|**uv;dv**)
≡ lim(vs(q+dq − qi)|**uv** − vs(q − qi)|**uv**)***qd**(**dv**)
= **0**, for all q

D[v − dv,dv]•(us(q − qi)|**uv;dv**)
≡ lim(us(q − qi)|**uv** − us(q − dq − qi)|**uv**)***qd**(**dv**)
= **0**, for all q

D[v − dv,dv]•(vs(q − qi)|**uv;dv**)
≡ lim(vs(q − qi)|**uv** − vs(q − dq − qi)|**uv**)•**qd**(**dv**)
= **0**, q ≠ qi
= **uv**/dq, q = qi

as dq→0

for

v(q) a curve over CQ;

then

the basic process gradient of a step function along CQ is defined as

D[v(q),**v**(q)+**dv**(q)]*(us(**v**(q) − **v**(qi))|CQ;**dv**(q))
≡ lim(us(**v**(q)+**dv**(q)− **v**(qi))|CQ −us(**v**(q) − **v**(qi))|CQ)
•**qd**(**dv**(**v**(qi)))
= **0**, **v** ≠ **v**(qi) in CQ
= **qd**(**dv**(**v**(qi))), **v**=**v**(qi)

D[v(q),**v**(q)+**dv**(q)]*(vs(**v**(q) − **v**(qi))|CQ;**dv**(q))
≡ lim(vs(**v**(q)+**dv**(q)− **vs**(qi))|CQ −v(**v**(q) − **v**(qi))|CQ)
•**qd**(**dv**(**v**(qi)))
= **0**, for all **v**(q) in CQ

D[v(q)-**dv**(q),**v**(q)]*(us(**v**(q) − **v**(qi))|CQ;**dv**(q))
≡ lim(us(**v**(qi)− **v**(q)-**dv**(q))|CQ −us(**v**(q) − **v**(qi))|CQ)
•**qd**(**dv**(**v**(qi)))
= **0**, for all **v**(q) in CQ

D[**v**(q)-**dv**(q),**v**(q)]∗(vs(**v**(q) - **v**(qi))|CQ;**dv**(q))
≡ lim(vs(**v**(qi)- **v**(q)-**dv**(q))|CQ -vs(**v**(q) - **v**(qi))|CQ)
•**qd**(**dv**(**v**(qi)))
= **0**, **v**(q)≠**v**(qi) in CQ
= **qd**(**dv**(**vi**)), **v**(q)=**v**(qi)
as **dv**(**v**(q))→**0**

for
v in CV a curve from **v1** to **v2**;
then
the basic process gradient of a step function along CV is defined as

D[**v**,**v**+**dv**(**v**)]∗(us(**v** - **vi**)|CV;**dv**)
≡ lim(us(**v**+**dv**(**v**)- **vi**)|CV -us(**v** - **vi**)|CV)∗**qd**(**dv**(**v**))
= **0**, **v** ≠ **vi** in CV
= **qd**(**dv**(**vi**)), **v**=**vi**

D[**v**,**v**+**dv**(**v**)]∗(vs(**v** - **vi**)|CV;**dv**)
≡ lim(vs(**v**+**dv**(**v**) - **vi**)|CV - vs(**v** - **vi**)|CV)•**qd**(**dv**(**v**))
= **0**, for all **v** in CV

D[**v** - **dv**(**v**),**v**]∗(us(**v** - **vi**)|CV;**dv**)
≡ lim(us(**vi** - **v** -**dv**(**v**))|CV - us(**v** - **vi**)|CV)∗**qd**(**dv**(**v**))
= **0**, for all **v** in CV

D[**v** - **dv**(**v**),**v**]∗(v(**v** - **vi**)|CV;**dv**)
≡ lim(vs(**vi** - **v**- **dv**(**v**))|CV - vs(**v** - **vi**)|CV)•**qd**(**dv**(**v**))
= **0**, **v**≠**vi** in CV
= **qd**(**dv**(**vi**)), **v**=**vi**
as **dv**(**v**)→**0**.
end of definition

Thus, like real impulse functions, these gradients are impulsive in amplitude but in addition they have a definite direction."

Newton: "The definition is similar to scalar derivatives along curves, differing only in that a scalar increment is replaced by a vector increment."

Breton: "Correct. So we may expect similar results."

Einstein, leading commandingly: "Now show us how these step function gradients interact with vector functions!"

Breton, obediently: "For **f** a generalized function over CV, consider

(**f**|CV)∗(**D**[**v**,**v**+**dv**]∗(us(**v**–**vi**)|CV);**dv**)
= **f**(**vi**)∗**qd**(**dv**(**vi**)), **v**=**vi**
= [**0**], otherwise;

(us(**v**–**vi**)|CV)∗**D**[**v**,**v**+**dv**]∗(**f**|CV;**dv**)
= [**0**], **v** in [**v0**,**vi**]
= **D**[**v**,**v**+**dv**]∗(**f**|CV;**dv**), otherwise;

and
D[**v**,**v+dv**]*((**f**|CV)*us(**v–vi**)|CV;**dv**)
= [**0**], **v** in [**v0**,**vi**)
= lim **f**(**vi+dv**(**vi**))***qd**(**dv**(**vi**)), **v**=**vi**
= **D**[**v**,**v+dv**]*(**f**|CV);**dv**), **v** in (**vi**,**v2**].

Newton: "These results flow directly from the definitions."

Breton: "If **f** is continuous at **vi** this last equation is compactly written as
D[**v**,**v+dv**]*((**f**|CV)*us(**v–vi**)|CV;**dv**)
= **f**(**vi**)***D**[**v**,**v+dv**](us(**v–vi**)|CV;**dv**)
+ vs(**v–vi**)|CV***D**[**v**,**v+dv**]*(**f**|CV;**dv**).

Einstein: "Where did you get that?"

Breton: "The derivative of the product can be split into three parts. For **v** in [**v0**,**vi**), the product is zero since us(**v–vi**) =0 there. For **v** in (**vi**,**v2**], the product is **D**[**v**,**v+dv**]*(**f**|CV;**dv**) since vs(**v–vi**) = 1 there. At **v=vi** the product becomes an impulse function with **f**(**vi**) as its vectorial value. The first expression that I wrote captures the first two parts; the second expression captures the third part."

Einstein: "Now won't you please draw us a diagram."

Breton, expansively: "Why not try yourself."

Einstein, pleased, complied with the request by drawing the following illustration."

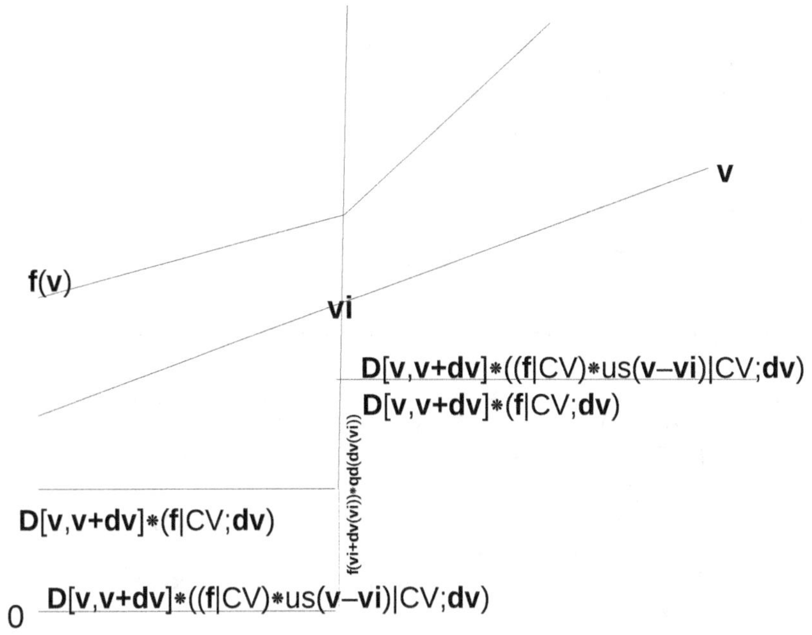

Einstein: "I'll take Breton's lead and simply illustrate in two dimensions and further simplify by taking a function which has only two constant derivatives, one before **vi** and the other after. Such a function is marked **f(v)** in the diagram. The constant derivatives are shown associated with the horizontal lines. The derivative of the product is shown as **0** on the bottom horizontal line which steps up to the line for **D[v,v+dv]*(f**|CV;**dv**) beyond **vi**."

Breton: "Remarkable is the action at **vi**. While it is easy to see the results prior to and beyond **vi**, the action at **vi** can easily be overlooked. For Theoretical Physics this action might be made to correspond to what happens at an interface."

Newton, happily filling out results for the other gradients: "Similarly,
D[v-dv,v]*(f|CV)*vs(**v-vi**)|CV);**dv**)
= **f(vi)*D[v-dv,v]**(vs(**v-vi**)|CV;**dv**)
+ vs(**v-vi**)|CV***D[v-dv,v]*(f**|CV;**dv**)
D[v,v+dv]*((f|uv)*us(**v-vi**)|**uv;dv**)
= **f(vi)*D[v,v+dv]***(us(**v-vi**)|**uv;dv**)
+ (vs(**v-vi**)|**uv**)***D[v,v+dv]*(f|uv;dv**)
D[v-dv,v]*((f|uv)*vs(**v-vi**)|**uv;dv**)
= **f(vi)*D[v-dv,v]***(vs(**v-vi**)|**uv;dv**)
+ (vs(**v-vi**)|**uv**)***D[v-dv,v]*(f|uv;dv**).

Einstein, following the now well blazed path: "How about the integrals?"

Breton, accepting the request expansively: "For **f** a generalized function over a curve,
I[CV]•((f|CV)∗D[v,v+dv]∗(us(v−vi)|CV;dv)|CV;dv)
= lim f(vi)∗qd(dv(vi))•dv(vi)
= f(vi).
Also,
I[CV]•((f|CV)∧D[v,v+dv]∗(us(v−vi)|CV;dv)|CV;dv)
= f(vi)∧qd(dv(vi))•dv(vi)
= f(vi)•qd(dv(vi))∧dv(vi)
= 0
I[CV]∧((f|CV)∗D[v,v+dv]∗(us(v−vi)|CV;dv)|CV;dv)
= 0
I[CV]∧((f|CV)∧D[v,v+dv]∗(us(v−vi)|CV;dv)|CV;dv)
= lim f(vi)•dv(vi)∗qd(dv(vi)) − f(vi)
I[CV]∗((f|CV)∗D[v,v+dv]∗(us(v−vi)|CV;dv)|CV;dv)
= lim f(vi)∗qd(dv(vi))∗dv(vi)
I[CV]∗((f|CV)∗D[v,v+dv]∧(u(v−vi)|CV;dv)|CV;dv)
= C(f(vi))•(qd(dv(vi))∗dv(vi))
I[CV]∗((f|CV)•D[v,v+dv]∧(us(v−vi)|CV;dv)|CV;dv)
= f(vi)•qd(dv(vi))∗dv(vi).
For vs(v−vi),
I[CV]•((f|CV)∗D[v−dv,v]∗(vs(v−vi)|CV;dv)|CV;dv)
= f(vi)
I[CV]•((f|CV)∧D[v−dv,v]∗(vs(v−vi)|CV;dv)|CV;dv)
= 0
I[CV]∧((f|CV)∗D[v−dv,v]∗(vs(v−vi)|CV;dv)|CV;dv)
= 0
I[CV]∧((f|CV)∧D[v−dv,v]∗(vs(v−vi)|CV;dv)|CV;dv)
= f(vi)•dv(vi)∗qd(dv(vi)) − f(vi)
I[CV]∗((f|CV)∗D[v−dv,v]∗(vs(v−vi)|CV;dv)|CV;dv)
= f(vi)∗qd(dvi)∗dvi
I[CV]∗((f|CV)∗D[v−dv,v]∧(vs(v−vi)|CV;dv)|CV;dv)
= C(f(vi))•(qd(dv(vi))∗dv(vi))
I[CV]∗((f|CV)•D[v−dv,v]∧(vs(v−vi)|CV;dv)|CV;dv)
= f(vi)•qd(dv(vi))∗dv(vi).

Newton, accepting the expansion: "Similar results hold for directional functions."

Einstein, also accepting: "Let's take the special cases."

Breton: "As you can see, the gradients allow an expansion of results involving divergences, curls and thus also results for invergences, incurls and ingradients.

Einstein: “So we will acquire many more valid equations. So much the better. Please continue.”

Breton: "Consider **f** bounded and continuous except possibly at **vi** interior to (**v0,vm**) where

lim **f**(**v**(i+1))|CV ≡ **f1**
lim **f**(**v**(i−1))|CV ≡ **f0**

Then

I[**v0**,**vm**]•((us(**v**−**vi**)|CV)∗**D**[**v**,**v**+**dv**(**v**)]∗(**f**(**v**)|CV;**dv**)|CV;**dv**)
= lim **D**[**v**(i+1),**v**(i+1)+**dv**(**v**(i+1))]∗(**f**(**v**)|CV;**dv**)
•**dv**(**v**(i+1))
+ ...
+ **D**[**vm**,**v**+**dv**(**vm**)]∗(**f**(**v**)|CV;**dv**)∗**dv**(**vm**)
= lim (**f**(**v**(i+2))|CV − **f**(**v**(i+1))|CV)
∗**qd**(**dv**(**v**(i+1)))•**dv**(**v**(i+1))
+ ...
+(**f**(**v**(m+1))|CV − **f**(**vm**)|CV)∗**qd**(**dv**(**vm**))
•**dv**(**vm**)
= lim − **f**(**v**(i+1))|CV) + (**f**(**v**(m+1))|CV
= **f**(**vm**) − **f1**.

Thus,

I[**v0**,**vm**]•((us(**v**−**vi**)|CV)∗**D**[**v**,**v**+**dv**(**v**)]∗(**f**(**v**)|CV;**dv**)|CV;**dv**)
= **f**(**vm**) − **f1**

I[**v0**,**vm**]•((us(**v**−**vi**)|CV)∗**D**[**v**,**v**+**dv**(**v**)]∧(**f**(**v**)|CV;**dv**)|CV;**dv**)
= 0

I[**v0**,**vm**]∧((us(**v**−**vi**)|CV)∗**D**[**v**,**v**+**dv**(**v**)]∧(**f**(**v**)|CV;**dv**)|CV;**dv**)
= **f**(**vm**)•[**dv**(**vm**)∗**qd**(**dv**(**vm**)) − **I**]
− **f1**•[**dv**(**v**(i+1))∗**qd**(**dv**(**v**(i+1))) − **I**]

I[**v0**,**vm**]∗((u(**v**−**vi**)|CV)∗**D**[**v**,**v**+**dv**(**v**)]∧(**f**(**v**)|CV;**dv**)|CV;**dv**)
= **C**(**f**(**vm**))•[**qd**(**dv**(**vm**))∗**dv**(**vm**)]
− **C**(**f1**)•[**qd**(**dv**(**v**(i+1)))∗**dv**(**v**(i+1))]

I[**v0**,**vm**]∗((us(**v**−**vi**)|CV)∗**D**[**v**,**v**+**dv**(**v**)]•(**f**(**v**)|CV;**dv**)|CV;**dv**)
= **f**(**vm**)•[**qd**(**dv**(**vm**))∗**dv**(**vm**)]
− **f1**•[**qd**(**dv**(**v**(i+1)))∗**dv**(**v**(i+1))].

Similarly,

I[**v0**,**vm**]•((vs(**v**−**vi**)|CV)∗**D**[**v**−**dv**(**v**),**v**]∗(**f**(**v**)|CV;**dv**)|CV;**dv**)
= **f**(**vm**) − **f0**

I[**v0**,**vm**]•((vs(**v**−**vi**)|CV)∗**D**[**v**−**dv**(**v**),**v**]∧(**f**(**v**)|CV;**dv**)|CV;**dv**)
= 0

I[**v0**,**vm**]∧((vs(**v**−**vi**)|CV)∗**D**[**v**−**dv**(**v**),**v**]∧(**f**(**v**)|CV;**dv**)|CV;**dv**)
= **f**(**vm**)•[**dv**(**vm**)∗**qd**(**dv**(**vm**)) − **I**]
− **f1**•[**dv**(**v**(i−1))∗**qd**(**dv**(**v**(i−1))) − **I**]

I[**v0**,**vm**]∗((vs(**v**−**vi**)|CV)∗**D**[**v**−**dv**(**v**),**v**]∧(**f**(**v**)|CV;**dv**)|CV;**dv**)
= **C**(**f**(**vm**))•[**qd**(**dv**(**vm**))∗**dv**(**vm**)]
− **C**(**f1**)•[**qd**(**dv**(**v**(i−1)))∗**dv**(**v**(i−1))]

I[v0,vm]∗((vs(v−vi)|CV)∗D[v−dv(v),v]•(f(v)|CV;dv)|CV;dv)
= f(vm)•[qd(dv(vm))∗dv(vm)]
− f1•[qd(dv(v(i−1)))∗dv(v(i−1))]
Similar results hold for directional functions."

Einstein, unrelenting: "How about the gradient of the product?"

Breton, complying:
I[v0,vm]•(D[v,v+dv(v)]∗(us(v−vi)|CV)∗(f(v)|CV;dv)|CV;dv)
= lim f(v(i+1))|CV∗qd(dv(vi))•dv(vi)
+ (f(v(i+2))|CV − f(v(i+1))|CV)
∗qd(dv(v(i+1)))•dv(v(i+1))
+ ...
+(f(v(m+1))|CV − f(vm)|CV)
∗qd(dv(vm))•dv(vm)
= lim f(v(m+1))|CV
= f(vm).
Thus,
I[v0,vm]•(D[v,v+dv(v)]∗(us(v−vi)|CV)∗(f(v)|CV;dv)|CV;dv)
= f(vm)
I[v0,vm]•(D[v,v+dv(v)]∗(us(v−vi)|CV)∧(f(v)|CV;dv)|CV;dv)
= 0
I[v0,vm]∧(D[v,v+dv(v)]∗(us(v−vi)|CV)∧(f(v)|CV;dv)|CV;dv)
= f(vm)•[dv(vm)∗qd(dv(vm)) − I]
I[v0,vm]∗(D[v,v+dv(v)]∗(us(v−vi)|CV)∧(f(v)|CV;dv)|CV;dv)
= C(f(vm))•[qd(dv(vm))∗dv(vm)]
I[v0,vm]∗(D[v,v+dv(v)]∗(us(v−vi)|CV)•(f(v)|CV;dv)|CV;dv)
= f(vm)•[qd(dv(vm))∗dv(vm)].
Similarly,
I[v0,vm]•(D[v−dv(v),v]∗(vs(v−vi)|CV)∗(f(v)|CV;dv)|CV;dv)
= f(vm)
I[v0,vm]•(D[v−dv(v),v]∗(vs(v−vi)|CV)∧(f(v)|CV;dv)|CV;dv)
= 0
I[v0,vm]∧(D[v−dv(v),v]∗(vs(v−vi)|CV)∧(f(v)|CV;dv)|CV;dv)
= f(vm)•[dv(vm)∗qd(dv(vm)) − I]
I[v0,vm]∗(D[v−dv(v),v]∗(vs(v−vi)|CV)∧(f(v)|CV;dv)|CV;dv)
= C(f(vm))•[qd(dv(vm))∗dv(vm)]
I[v0,vm]∗(D[v−dv(v),v]∗(vs(v−vi)|CV)•(f(v)|CV;dv)|CV;dv)
= f(vm)•[qd(dv(vm))∗dv(vm)].

Newton: "Consequently for f continuous at vi,
I[v0,vm]•(D[v,v+dv(v)]∗(us(v−vi)|CV)∗(f(v)|CV;dv)|CV;dv)
= I[v0,vm]•((us(v−vi)|CV)∗D[v,v+dv(v)]∗(f(v)
|CV;dv)|CV;dv)
+ I[v0,vm]•((f|CV)∗D[v,v+dv]∗(u(v−vi)
|CV;dv);dv).

Breton: "Thus, the integral product rule holds only selectively for these many invergences, incurls, and ingradients over curved lines."

Newton: "These many equations address many subtleties."

Breton, continuing: "Now let C1 = [**v0**,**vi**) and C2 = [**vi**,**vm**]. Then for f=1

$$\mathtt{I}[C1]\bullet(\mathbf{D}[\mathbf{v},\mathbf{v}+\mathbf{dv}(\mathbf{v})]*(us(\mathbf{v}-\mathbf{vi})\,|\,CV;\mathbf{dv})\,|\,CV;\mathbf{dv}) = 0$$

while

$$\mathtt{I}[C2]\bullet(\mathbf{D}[\mathbf{v},\mathbf{v}+\mathbf{dv}(\mathbf{v})]*(us(\mathbf{v}-\mathbf{vi})\,|\,CV;\mathbf{dv})\,|\,CV;\mathbf{dv}) = 1.$$

Similarly, for C1 = [**v0**,**vi**) and C2 = [**vi**,**vm**]

$$\mathtt{I}[C1]\bullet(\mathbf{D}[\mathbf{v},\mathbf{v}+\mathbf{dv}(\mathbf{v})]*(vs(\mathbf{v}-\mathbf{vi})\,|\,CV;\mathbf{dv})\,|\,CV;\mathbf{dv}) = 0$$

while

$$\mathtt{I}[C2]\bullet(\mathbf{D}[\mathbf{v},\mathbf{v}+\mathbf{dv}(\mathbf{v})]*(vs(\mathbf{v}-\mathbf{vi})\,|\,CV;\mathbf{dv})\,|\,CV;\mathbf{dv}) = 1.$$

Newton: "Interesting."

Breton: "Now let

$$C1 \equiv [\mathbf{v1},\mathbf{v2}] \subset CV \equiv [\mathbf{v0},\mathbf{vm2}].$$

Then

$$\mathtt{I}[C1]\bullet(\mathbf{f}\,|\,CV;\mathbf{dv}) = \mathtt{I}[CV]\bullet(vs(\mathbf{v}-\mathbf{v1})-us(\mathbf{v}-\mathbf{v2}))\bullet(\mathbf{f}(\mathbf{v})\,|\,CV);\mathbf{dv}).$$

In this way even though **f** may be unbounded on some portions of a curve, it may be still be integrated over portions over which it is bounded."

Newton: "A result which can prove useful indeed."

Breton, still continuing: "Now for **d** a vector constant let

$$\mathbf{f}(\mathbf{v})\,|\,CV = \mathbf{f1}(\mathbf{v})\,|\,CV + \mathbf{d}\bullet us(\mathbf{v}-\mathbf{vi})$$

where **f1** is bounded and continuous over CV such that

$$\mathtt{I}[CV]\bullet(\mathbf{f1}(\mathbf{v})\,|\,CV;\mathbf{dv}) = \mathbf{G}(\mathbf{vm}) - \mathbf{G}(\mathbf{v0}).$$

Then

$$\begin{aligned}\mathtt{I}[CV]\bullet(\mathbf{f}\,|\,CV;\mathbf{dv}) &\\ &= \mathtt{I}[CV]\bullet(\mathbf{f1}\,|\,CV;\mathbf{dv}) + \mathbf{d}\bullet\mathtt{I}[\mathbf{vi},\mathbf{vm}]\bullet(1\,|\,CV;\mathbf{dv})\\ &= \mathbf{G}(\mathbf{vm}) - \mathbf{G}(\mathbf{v0}) + \mathbf{d}\bullet\mathtt{I}[\mathbf{vi},\mathbf{vm}]\bullet(1\,|\,CV;\mathbf{dv}).\end{aligned}$$

Similarly for

$$\mathbf{f}(\mathbf{v})\,|\,CV = \mathbf{f1}(\mathbf{v})\,|\,CV + \mathbf{d}\bullet vs(\mathbf{v}-\mathbf{vi})$$

$$\mathtt{I}[CV]\bullet(\mathbf{f}\,|\,CV;\mathbf{dv}) = \mathbf{G}(\mathbf{vm}) - \mathbf{G}(\mathbf{v0}) + \mathbf{d}\bullet\mathtt{I}[\mathbf{vi},\mathbf{vm}]\bullet(1\,|\,CV;\mathbf{dv}).$$

These equations are also **extensions to the fundamental theorem of integral calculus**.

Analogous conclusions apply to invergences and incurls. These results further expand the fundamental theorem of integral calculus to functions with a finite number of finite discontinuities in a given direction or over curves."

Einstein: "How about derivatives over regions?"

The following development establishes a theory of local integration which inverts sectional gradients.

Vector measures in V3

Breton: "Similar to integration over Q, integration over **V3** involves partitioning and measuring.

Definition (partitions of V3)

Given

p1, p2, p3, positive quotient numbers;
u1V1, **u2V1**, **u3V1**, non-parallel directions;

for

n>0, a given integer;
i,j,k= – (n^2-1),...,–,1,0,1,2,3,...n^2;
$(i-1)/n \le t1 < i/n$ for abs(t1)≤n;
$(j-1)/n \le t2 < j/n$ for abs(t2)≤n;
$(k-1)/n \le t3 < k/n$ for abs(t3)≤n;

Pn ≡ {**v**=t1*p1***u1V1** + t2*p2***u2V1** + t3*p3***u3V1**}
for all t1, t2, and t3 with Pn_∞ otherwise.

end of definition

Einstein: "Can you illustrate this?"

Breton: "Try this.

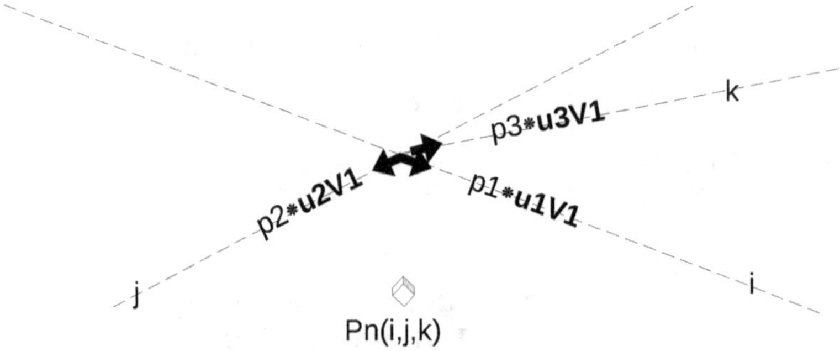

The vectors p1***u1V1**, p2***u2V1**, p3***u3V1** are shown in short, dark lines with arrows. The indicated axes are marked by corresponding indices: i,j,k. When the vectors are multiplied by a given n, the axes are extended to mark a vicinity of V3. The vicinity is diced into so many little cells by the indices, one for each separate combination. One such cell is shown as Pn(i,j,k). The entire set of such cells is called a partition of the defined vicinity."

Newton: "Why is it called a partition?"

Breton: "Any **v** of V3 may be placed into one and only one of the subsets of Pn. Pn is thus a partition of V3 into $(2*n^2)^3 + 1$ disjoint subsets Pn(i,j,k) called **cells**."

Newton: "Each of the cells is the same."

Breton: "Not the same, but rather just a translated image of any other cell."

Einstein, summarizing: "So we end up with a defined vicinity, and a finite number of similar cells."

Breton: "For a given n. What happens as n increases?"

Einstein, answering loftily: "The vicinity comprehended by the partition grows larger and larger."

Breton: "How about the size of any cell?"

Einstein: "Don't the cells remain the same?"

Breton: "Let's look at one edge of a cell as a function of n. From the definition: $((i-1)/n) \le t1 < (i/n)$. The width of t1 then is just

$$(i/n) - ((i-1)/n) = (i - i+1)/n = 1/n$$

So the ith side of the cell is shrinking, and so the other two as well."

Einstein, resummarizing: "So with increasing n, the partition becomes becomes larger that is, encompasses more and more cells of smaller and smaller size."

Breton: "So by multiplying the increasing number of cells with a decreasing cellular size, we could establish a condition which encourages the possibility of limits."

Newton, seeing similarities: "Similar to the definition of other integrals."

Einstein: "Cells in a partition may look the same but each has a different location."

Breton: "Each cell occupies many locations, any of which might be used to locate the cell. For the sake of uniformity, let us choose the center of the cell as its location. Here is the definition.

Definition (location of a cell in a partition)

Given

Pn(i,j,k), a cell in partition Pn;

then

$$\mathbf{v}(Pn(i,j,k)) \equiv (i - ½)*p1*\mathbf{u1V1}/n + (j - ½)*p2*\mathbf{u2V1}/n + (k - ½)*p3*\mathbf{u3V1}/n$$

$$\mathbf{v}(Pn\infty) \equiv 0.$$

end of definition

Einstein: "So we have partitioned **V3**. What next?"

Breton: "Next we need to establish a measure for the integration. For each Pn(i,j,k) of Pn let

$$\mathbf{mn}(Pn(i,j,k)) \equiv ((\mathbf{v}(Pn(i+1,j,k)) - \mathbf{v}(Pn(i,j,k)))*\mathbf{u1V1} + (\mathbf{v}(Pn(i,j+1,k)) - \mathbf{v}(Pn(i,j,k)))*\mathbf{u2V1} + (\mathbf{v}(Pn(i,j,k+1)) - \mathbf{v}(Pn(i,j,k)))*\mathbf{u3V1})/n^2 = (p1*\mathbf{u1V1}+p2*\mathbf{u2V1}+p3*\mathbf{u3V1})/n^3$$

with $\mathbf{mn}(Pn\infty) \equiv \mathbf{0}$."

Newton, commenting: "Then each cell, with the exception of Pn∞, has the same vectorial value."

Einstein: "And how does that lead to integration?"

Breton: "For a measurable set M of **V3** in a given partition Pn, let bn(M∩Pn) be the number of cells for which M∩Pn(i,j,k) is non-empty. Then define the **measure of M with respect to the partition Pn** as

$$\mathbf{mn}(M \cap Pn) \equiv bn(M \cap Pn)*(\mathbf{mn}(Pn(i,j,k))) = (p1*\mathbf{u1V1}+p2*\mathbf{u2V1}+p3*\mathbf{u3V1})*bn(M \cap Pn)/n^3.$$

Einstein, concluding: "So we have a different measure for the same measurable set depending on n. That does not appear hopeful."

Breton: "Patience my dear Einstein. Here is the definition of the measure of a measurable set of vectors.

Definition (measure of a measurable set M)
Given
$\mathbf{pV1} = (p1*\mathbf{u1V1}+p2*\mathbf{u2V1}+p3*\mathbf{u3V1})$;
M a measurable set of **V3**;
{Pn}, a sequenced set of partitions of **V3** based on **pV1**;
$\mathbf{mn}(M \cap Pn)$, the measure of M with respect to the partition Pn;
then

$$\mathbf{m}(M) \equiv \lim(\mathbf{mn}(M \cap Pn)) \text{ as } n \to \infty$$
$$\equiv m(M)*\mathbf{pV1}.$$

end of definition

Newton, reflecting: "The measure of M is thus seen to consist of a scalar

$$m(M) = \lim\ (1/n^3)*bn(M \cap Pn) \text{ as } n \to \infty$$

and a vector,

$$\mathbf{pV1} \equiv p1*\mathbf{u1V1}+p2*\mathbf{u2V1}+p3*\mathbf{u3V1},$$

called the **partition vector**."

Breton: "The measure of a given measurable set in **V3** depends on the partition vector. Where the dependence needs to be made explicit, the measure of M is written as a condition:

$$\mathbf{m}(M|\mathbf{pV1}) = m(M|\mathbf{pV1})*\mathbf{pV1}.$$

It is instructive to examine this measure in terms of its associated volumes."

Theorem (Measure and volumes)

Given

M, a measurable set of V3;

vol(M), volume of M;

$\mathbf{pV1} \equiv p1\ast\mathbf{u1V1}+p\,2\ast\mathbf{u2V1}+p3\ast\mathbf{u3V1}$, a partition vector;

for

$\mathbf{p} \equiv p1\ast\mathbf{u1} + p2\ast\mathbf{u2} + p3\ast\mathbf{u3}$;

$\mathbf{G(p)}$, the diagonal vector operator

$\mathbf{UV1} \equiv \mathbf{u1}\ast\mathbf{u1V1} + \mathbf{u2}\ast\mathbf{u2V1} + \mathbf{u3}\ast\mathbf{u3V1}$;

then

$$\begin{aligned}
\mathrm{vol}(Pn(i,j,k)) &= \det[\mathbf{G(p)}]\ast\det[\mathbf{UV1}]/n^3 \\
&= \mathrm{vol}(P1(i,j,k))/n^3 \\
\mathbf{mn}(Pn(i,j,k)) &= \mathrm{vol}(Pn(i,j,k))\ast\mathbf{pV1}/(\det[\mathbf{G(p)}]\ast\det[\mathbf{UV1}]) \\
&= \mathrm{vol}(Pn(i,j,k))\ast\mathbf{pV1}/\mathrm{vol}(P1(i,j,k)) \\
\mathbf{mn}(M\,|\,\mathbf{pV1}) &= bn(M\cap Pn)\ast\mathrm{vol}(Pn(i,j,k))\ast\mathbf{pV1} \\
&\qquad /(\det[\mathbf{G(p)}]\ast\det[\mathbf{UV1}]) \\
&= bn(M\cap Pn)\ast\mathrm{vol}(Pn(i,j,k))\ast\mathbf{pV1}/\mathrm{vol}(P1(i,j,k)) \\
\mathbf{m}(M\,|\,\mathbf{pV1}) &= \mathrm{vol}(M)\ast\mathbf{pV1}/\mathrm{vol}(P1(i,j,k)).
\end{aligned}$$

Proof:

$$\begin{aligned}
\mathrm{vol}(P1(i,j,k)) &= (p1\ast\mathbf{u1V1})\bullet(p2\ast\mathbf{u2V1})\wedge(p3\ast\mathbf{u3V1}) \\
&= p1\ast p2\ast p3\ast(\mathbf{u1V1}\bullet((\mathbf{u2V1})\wedge(p3\ast\mathbf{u3V1}))) \\
&= \det[\mathbf{G(p)}]\ast\det[\mathbf{UV1}].
\end{aligned}$$

Consequently,

$$\mathrm{vol}(Pn(i,j,k)) = \det[\mathbf{G(p)}]\ast\det[\mathbf{UV1}]/n^3$$

Again,

$$\begin{aligned}
mn(Pn(i,j,k)) &= 1/n^3 \\
&= \mathrm{vol}(Pn(i,j,k))/(\det[\mathbf{G(p)}]\ast\det[\mathbf{UV1}]) \\
&= \mathrm{vol}(Pn(i,j,k))/\mathrm{vol}(P1(i,j,k)) \\
mn(M\cap Pn) &= bn(M\cap Pn)/n^3 \\
&= bn(M\cap Pn)\ast\mathrm{vol}(Pn(i,j,k))/\mathrm{vol}(P1(i,j,k))
\end{aligned}$$

Finally for a measurable set

$$\lim bn(M\cap Pn)\ast\mathrm{vol}(Pn(i,j,k)) = \mathrm{vol}(M).$$

qed

Einstein: "This is interesting indeed. Would you illustrate these ideas?"

Breton: "Let me illustrate in two dimensions only. The first shows the measurable set of vectors in a coarse mesh, that is, in some partition Pn. The second shows the same set as the mesh becomes finer, that is, in a partition of greater n."

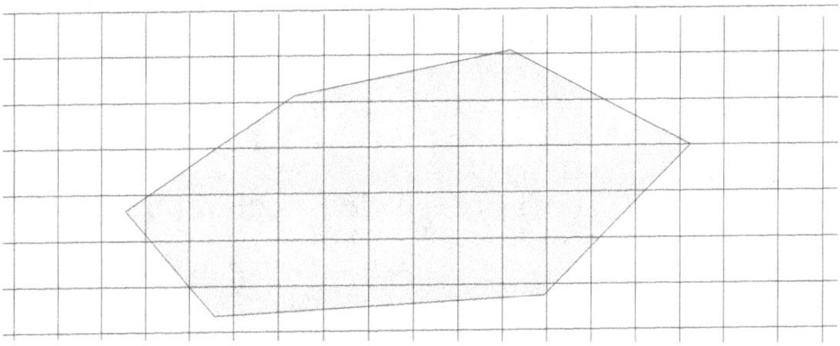

Coarse Mesh

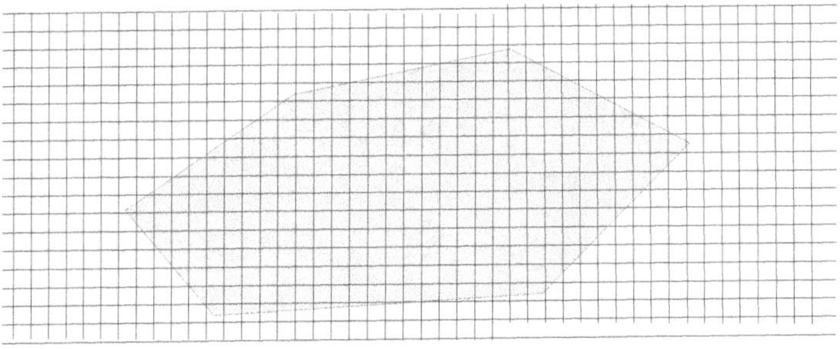

Finer Mesh

Newton, repeating: "With the finer mesh the number or cells increases, as well as the cells which intersect with the set."

Breton: "Balanced by a reduction in the size of the cells. Notice with the finer mesh how cells which intersect the surface of the set grows closer to that surface."

Einstein. objecting: "What is the surface? You have not defined it."

Breton: "You're right, I have not. Here is the somewhat intricate definition.

Definition 2 (Surface of M)

Given

pV1 = (p1***u1V1**+p2***u2V1**+p3***u3V1**);
M, a measurable set of V3;
{Pn}, a sequenced set of partitions of V3 based on **pV1**;
mn(M∩Pn), the measure of M with respect to the partition Pn;

then

if Pn(i,j,k) ∩ M = Pn(i,j,k)
Pn(i,j,k) is called a cell **interior to** M

if Pn(i,j,k) ∩ M = null
Pn(i,j,k) is called a cell **exterior to** M

if Pn(i,j,k) ∩ M ⊂ Pn(i,j,k) properly
Pn(i,j,k) is called a **surface cell of** M.

The set
SM ≡ {lim **v**(Pn(i,j,k)) | Pn(i,j,k) is a surface cell} as n→∞
is called the **surface of** M."

end of definition

Einstein: "Please illustrate."

Breton: "This sketch may help.

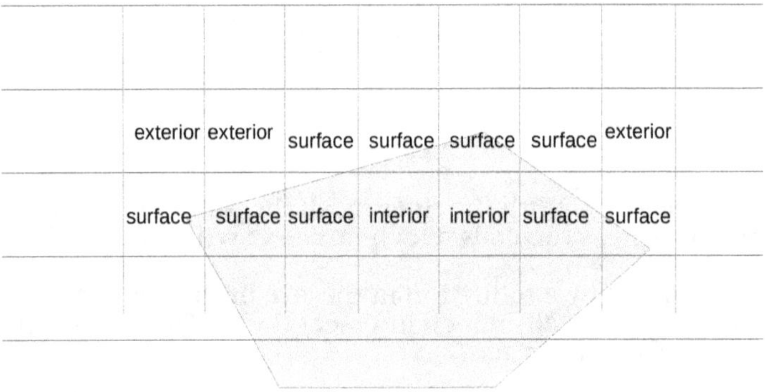

Given a partition Pn and a measurable set M, the intersection of M with the cells of the partition can be grouped into three classes: interior, exterior, or surface. Each surface cell contains vectors of both in M and those not in M.

As the mesh becomes finer with increasing n, the surface cells approximate the surface of M more accurately."

Newton: "This looks promising for dealing with limits."

Breton: "The surface of M may be further distinguished:
for **pV1** = pV1***upV1** for any pV1>0
SMPE|**pV1** ≡ {**v** in SM |(**v+pV1**/n^3 is in V3 - (M∪SM)
is called the **exterior positive surface of** M
SMPI|**pV1** ≡ {**v** in SM |(**v+pV1**/n^3 is in (M - SM)
is called the **interior positive surface of** M
SMP0|**pV1** ≡ {**v** in SM |(**v+pV1**/n^3 is in SM
is called the **zero positive surface of** M

Einstein: "An illustration would help."

Breton: "Here it is.

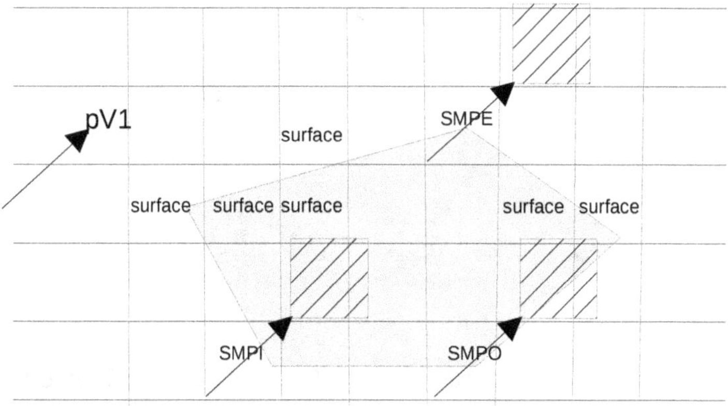

Breton: "Again only two dimensions are shown. Given the partition vector **pV1**, a trial can be made of each surface cell. If addition of each vector in the cell with pV1 produces another cell wholly exterior to M, the cell surface cell is designated as exterior, SMPE"

Einstein. objecting: "In the diagram only one such addition is shown."

Newton, replying: "The partition vector acts as a translation vector. The illustration shows the translation of one vector and the result of translation all the vectors in the cell."

Breton: "Thanks for stating my intention so nicely. If the location of the translated cell falls completely within M, the surface cell is labeled interior (SMPI). If the location of the translated cell falls on another surface cell, the surface cell is labeled the zero positive surface (SMPO)."

Newton, reflecting: "With increasing n not only do the surface cells change, but so do SMPE, SMPI, and SMPO."

Einstein: "Why not the negative of the partition vector?"

Breton: "Why not indeed? Here are some definitions.
SMNE|**pV1** ≡ {**v** in SM |(**v**–**pV1**/n^3 is in V3 – (M∪SM)
is called the **exterior negative surface of** M
SMNI|**pV1** ≡ {**v** in SM |(**v**–**pV1**/n^3 is in (M – SM)
is called the **interior negative surface of** M
SMN0|**pV1** ≡ {**v** in SM |(**v**–**pV1**/n^3 is in SM
is called the **zero negative surface of** M.
and an illustration."

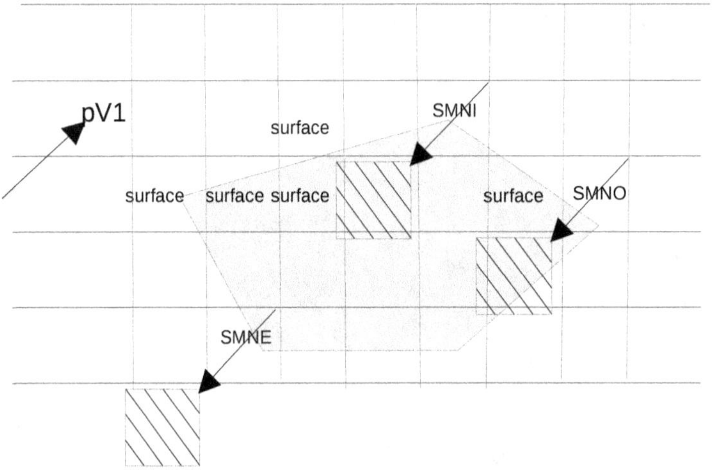

Newton, concluding: "So the same surface cells can be labeled differently."

Breton: "The sets SMPE|**pV1**, SMPI|**pV1**, and SMP0|**pV1**, partition SM as do the sets SMNE|**pV1**, SMNI|**pV1**, and SMN0|**pV1**. These are two different partitions of the same set of surface cells."

Einstein, commandingly: "Have you anything else to say?"

Breton: "So much more. Let us continue with measurable sets. For M1 and M2 measurable and disjoint,
m(M1UM2) = **m**(M1) + **m**(M2)
= (m(M1) + m(M2))∗**pV1**
Also,
m(Pn(i,j,k)) = **mn**(Pn(i,j,k)).

Einstein: "That's a startling assertion. Prove it!"

Breton: "Let's start with a definite cell, (Pn(i,j,k)) in the Pn partition.
The measure of that cell in that partition is
mn(Pn(i,j,k)) = (p1∗**u1V1**+p2∗**u2V1**+p3∗**u3V1**)/n^3
Now switch to the P2n partition. What is the measure of Pn(i,j,k) in the P2n partition?"

Newton: "Each cell in the new partition has a measure

$$\begin{aligned}\mathbf{m2n}(P2n(i,j,k)) \equiv\ & ((\mathbf{v}(Pn(i+1,j,k)) - \mathbf{v}(Pn(i,j,k)))*\mathbf{u1V1} \\ & + (\mathbf{v}(Pn(i,j+1,k)) - \mathbf{v}(Pn(i,j,k)))*\mathbf{u2V1} \\ & + (\mathbf{v}(Pn(i,j,k+1)) - \mathbf{v}(Pn(i,j,k)))*\mathbf{u3V1})/n^2 \\ =\ & (p1*\mathbf{u1V1}+p2*\mathbf{u2V1}+p3*\mathbf{u3V1})/(2*n)^3.\end{aligned}$$

Breton: "How many of the P2n(i,j,k) cells fit into one Pn(i,j,k) cell?

Newton: "Each P2n(i,j,k) is small enough to fit into a Pn(i,j,k) cell. In fact each defining lengths of the P2n(i,j,k) cell is just half of those for Pn(i,j,k)."

Breton: "The answer can be more easily seen by comparing two cubes. Inside a given cube a total of eight smaller cubes can fit each of which has a defining length one-half of the larger cube."

Newton: "So

$$\begin{aligned}\mathbf{m2n}(Pn(i,j,k)) &= 8*(p1*\mathbf{u1V1}+p2*\mathbf{u2V1}+p3*\mathbf{u3V1})/(2*n)^3 \\ &= (p1*\mathbf{u1V1}+p2*\mathbf{u2V1}+p3*\mathbf{u3V1})/n^3 \\ &= \mathbf{mn}(Pn(i,j,k)).\end{aligned}$$

Consequently the limit, which exists for measurable sets, is established by the sub-sequence **mn**, **m2n**, **m4n**, **m8n**,..."

Einstein, impressed: "Remarkable. Have you more to say?"

Breton: "The set M has a definite location. Suppose it is moved to a different location. Has its measure changed?"

Newton: "I don't think so."

Breton: "You are correct. Here is theorem to prove it."

Theorem (Measure of a translated set)

Given

a = a1***u1** + a2***u2** + a3***u3**, a translation vector;

M, a set with measure

m(M) = m(M)***pR1**;

for

MA = {**a** + **v**|**v** in M}

then

$$\mathbf{m}(MA) = \mathbf{m}(M).$$

Proof:

If M is measurable, then MA is also measurable. If {Pn} is a nested sequence of partitions of V3, then so also is {**a**+Pn}. Moreover, **m**(M) is unchanged in {**a**+Pn}. Now **m**(MA) = **m**(M) in {**a**+Pn}. So also then in {Pn}.

qed

Einstein: "Anything else?"

Breton: "The following theorem relates measures to different partitions, enabling partitions to be chosen conveniently."

Theorem (Measures with different partition vectors)

Given

pV1 ≡ p1V1***u1V1**+p2V1***u2V1**+p3V1***u3V1**, a partition vector;

M, a measurable set;

m(M|**pV1**) ≡ m(M|**pV1**)***pV1**;

pV2 ≡ p1V2***u1V2**+p2V2***u2V2**+p3V2***u3V2**, a second partition vector;

p1 ≡ p1V1***u1**+p2V1***u2**+p3V1***u3**;

p2 ≡ p1V2***u1**+p2V2***u2**+p3V2***u3**;

UV1 = u1*u1V1 + u2*u2V1 + u3*u3V1;

UV2 = u1*u1V2 + u2*u2V2 + u3*u3V2;

then

m(M|**pV2**) = m(M|**pV1**)
*det[**UV1**]*det[**G**(**p1**)]/(det[**UV2**]*det[**G**(**p2**)]).

Proof:
m(M|**pV1**) = vol(M)/(det[**UV1**]*det[**G**(**p1**)])
m(M|**pV2**) = vol(M)/(det[**UV2**]*det[**G**(**p2**)]).
Consequently,
m(M|**pV1**)*det[**UV1**]*det[**G**(**p1**)]
= m(M|**pV2**)*det[**UV2**]*det[**G**(**p2**)]
= vol(M). qed

Einstein: "What has all this to do with integration?"

Integration over Measurable Sets

Breton: "Everything, as you can see in the following definition."

Definition (Invergences, incurls, and ingradients over **V3**)

Given

f, a measurable function over **V3**;

M, a measurable set of **V3**;

pV1 = pV1***uV1**
= p1***u1V1**+p2***u2V1**+p3***u3V1**, a partition vector;

for

S[bn(M∩Pn)], the sum over which M∩Pn is non-empty

I[M]•(**f**(**r**)|**m**;**dV1**)
≡ lim S[bn(M∩Pn)](**f**(**r**(Pn(i,j,k)))•**m**(Pn(i,j,k)))
= lim S[bn(M∩Pn)](**f**(**r**(Pn(i,j,k)))•**pV1**/n^3)

I[M]∧(**f**(**r**)|**m**;**dV1**)
≡ lim S[bn(M∩Pn)](**f**(**r**(Pn(i,j,k)))∧**m**(Pn(i,j,k)))
= lim S[bn(M∩Pn)](**f**(**r**(Pn(i,j,k)))∧**pV1**/n^3)

I[M]∗(f(r)|m;dV1)
≡ lim S[bn(M∩Pn)](f(r(Pn(i,j,k)))∗m(Pn(i,j,k)))
= lim S[bn(M∩Pn)](f(r(Pn(i,j,k)))∗pV1/n^3)
as n→∞.
end of definition

Einstein: "These integrals are beginning to look like classical integrals."

Breton: "I can make them look even more so. For simplicity define

dV1 ≡ pV1/n^3

Then the measure of M can be written as

m(M) = m(M)∗pV1
= lim bn(M∩Pn)∗(pV1/n^3)
= lim bn(M∩Pn)∗dV1.

and the invergence, incurl, and ingradient may be written:

I[M]•(f(v)|m;dV1) ≡ lim S[bn(M∩Pn)](f(v(Pn(i,j,k)))•dV1)
I[M]∧(f(v)|m;dV1) ≡ lim S[bn(M∩Pn)](f(v(Pn(i,j,k)))∧dV1)
I[M]∗(f(v)|m;dV1) ≡ lim S[bn(M∩Pn)](f(v(Pn(i,j,k)))∗dV1)
as n→∞.

Newton: "Clear enough."

Breton: "We can go further. For abs(m(Pn(i,j,k))):V3→Q

I[M](f(v)|m;dV1)
≡ lim (S[bn(M∩Pn)](f(v(Pn(i,j,k)))∗abs(m(Pn(i,j,k))))
= lim S[bn(M∩Pn)](f(v(Pn(i,j,k)))/n^3)
∗sqrt((pV1•u1)^2 +(pV1•u2)^2 +(pV1•u3)^2).

Since

lim S[bn(M∩Pn)](f(v(Pn(i,j,k)))•pV1/n^3)
= lim S[bn(M∩Pn)](f(v(Pn(i,j,k))/n^3))•pV1

invergences, incurls and ingradients may be written as:

I[M]•(f(v)|m;dV1) ≡ I[M](f(v)|m)•pV1
I[M]∧(f(v)|m;dV1) ≡ I[M](f(v)|m)∧pV1
I[M]∗(f(v)|m;dV1) ≡ IM](f(v)|m)∗pV1

where

I[M](f(v)|m) ≡ lim S[bn(M∩Pn)](f(v(Pn(i,j,k))/n^3)).

Einstein: "The flexibility of the notation is commendable."

Breton: "You may be interested in the fact that integrals may also be defined in terms of functions of the partition vector **pV1** as in

I[M]•(f(v)|a;C(dV1))
≡ lim S[bn(M∩Pn)](f(v(Pn(i,j,k)))•C(pV1)/n^3)
= lim S[bn(M∩Pn)](f(v(Pn(i,j,k)))•C(dV1))

where **C** is the curl vector operator

or, for another example, as

$$\mathrm{I}[\mathrm{M}]\bullet(\mathbf{f}(\mathbf{v})|\mathrm{a};\mathbf{dV1}\bullet\mathbf{UV1}^{-1})$$
$$\equiv \lim \mathrm{S}[\mathrm{bn}(\mathrm{M}\cap\mathrm{Pn})](\mathbf{f}(\mathbf{v}(\mathrm{Pn}(i,j,k)))\bullet(\mathbf{pV1}\cdot\mathbf{UV1}^{-1})/n^{3})$$
$$= \lim \mathrm{S}[\mathrm{bn}(\mathrm{M}\cap\mathrm{Pn})](\mathbf{f}(\mathbf{v}(\mathrm{Pn}(i,j,k)))\bullet(\mathbf{dV1}\cdot\mathbf{UV1}^{-1})).$$

Einstein: "How about integrals using different partition vectors?"

Breton: "Let's start simply."

Theorem (relationships between integral s with different partition vectors)

Given

M a measurable set in **V3**;

f a continuous function over M;

pV1 = p1V1***u1V1** + p2V1***u2V1** + p3V1***u3V1**, a partition vector;

I[M]*(**f**;**dV1**) ≡ **I**[M](**f**(**v**)|**m1**)***pV1**;

for

pV2 = p1V2***u1V2** + p2V2***u2V2** + p3V2***u3V2**,
a second partition vector;

p1 ≡ p1V1***u1** + p2V1***u2** + p3V1***u3**;

p2 ≡ p1V2***u1** + p2V2***u2** + p3V2***u3**;

UV1 ≡ **u1*u1V1 + u2*u2V1 + u3*u3V1**;

UV2 ≡ **u1*u1V2 + u2*u2V2 + u3*u3V2**;

then

I[M]•(f;**dV2**) =I[M](**f**(**v**)|**m1**)•**pV1***det[G(**p1**)]*det[**UV1**]
/(det[G(**p2**)*det[**UV2**])

I[M]∧(**f**;**dV2**) =**I**[M](**f**(**v**)|**m1**)∧**pV1***det[G(**p1**)]*det[**UV1**]
/(det[G(**p2**)]*det[**UV2**])

I[M]*(**f**;**dV2**) =**I**[M](**f**(**v**)|**m1**)***pV1***det[G(**p1**)]*det[**UV1**]
/(det[G(**p2**)]*det[**UV2**]).

Proof:

Designate the n1 partition under **pV1** as PV1n1;
designate the n2 partition under **pV2** as PV2n2.

mn1(M∩Pn1|**pV1**) = bn1(M∩Pn1|**pV1**)/n1^3

mn2(M∩Pn2|**pV2**) = bn2(M∩Pn2|**pV2**)/n2^3

By the dilation theorem

m(PVn1(i,j,k)|**pV2**) = m(PV1n(i,j,k)|**pV1**)*det[**UV1**]*det[**G**(**p1**)]
/(det[**UV2**]*det[**G**(**p2**)])

For n1=n2

bn2(PVn1(i,j,k)∩Pn2) = det[**UV1**]*det[**G**(**p1**)]/(det[**UV2**]*det[**G**(**p2**)])

Thus one cell under the PV1n1 partition is occupied by

det[**UV1**]*det[**G**(**p1**)]/(det[**UV2**]*det[**G**(**p2**)])

cells under the PV2n1 partition.

Now adjust the ration n2/n1 such that

bn2(PVn1(i,j,k)∩Pn2) ≈ 1.

For such a ratio and number of cells in the PV2n2 partition equals the number of cells in the PV1n1 partition. Thus there exists a 1-1 correspondence between **v**(Pn1(i,,j,k)) and **v**(Pn2(i',j',k')).

Further

(n1/n2)^3 = det[**UV1**]*det[**G**(**p1**)]/(det[**UV2**]*det[**G**(**p2**)])

bn1(M∩Pn1)] ≈ bn2(M∩Pn2)].

Moreover, with the ratio n2/n1 maintained

v(Pn2(i',j',k')) □ **v**(Pn1(i,j,k)) as n1 ⟶ ∞

and since **f** is continuous,

f(**v**(PV2n2(i',j',k'))) ⟶ **f**(**v**(PV1n1(i,j,k))).

Consequently,
lim S[bn1(M∩Pn1)](**f**(**v**(Pn(i,j,k)))/$n1^3$)
= lim S[bn2(M∩Pn2)](**f**(**v**(Pn(i',j',k')))/$n2^3$)
*det[**UV2**]*det[**G**(**p2**)]/(det[**UV1**]*det[**G**(**p1**)]).
Consequently,
I[M]•(**f**;**dV2**) = I[M](**f**(**r**)|**m1**)•**pV2**
*det[G(**p1**)]*det[**UV1**]/(det[G(**p2**)]*det[**UV2**])
I[M]∧(**f**;**dV2**) = I[M](**f**(**r**)|**m1**)∧**pV2**}
*det[G(**p1**)]*det[**UV1**]/(det[G(**p2**)]*det[**UV2**])
I[M]*(**f**;**dV2**) = I[M](**f**(**r**)|**m1**)***pV2**}
*det[G(**p1**)]*det[**UV1**]/(det[G(**p2**)]*det[**UV2**]).
qed

Newton: "So the integrals of functions just follow the measures."

Breton: "Although a specific partition vector must be referenced for a proper definition, the following generic properties of these integrals are true for any given partition vector, **pV1**.

If M is the intersection of all open sets containing **v1** and **f** is bounded at **v1**

I[**v1**]•(**f**|**m**;**dV1**) = lim (S[bn(**v1**∩Pn)](**f**•**m**(Pn(i,j,k))))
= 0
I[**v1**]∧(**f**|**m**;**dV1**) = lim (S[bn(**v1**∩Pn)](**f**∧**m**(Pn(i,j,k))))
= **0**
I[**v1**]*(**f**|**m**;**dV1**) = lim (S[bn(**v1**∩Pn)](**f*****m**(Pn(i,j,k))))
= [**0**].

If **c** is a constant vector, for any measurable set M

I[M]•(**c**|**m**;**dV1**) = lim (S[bn(M∩Pn)](**c**•**m**(Pn(i,j,k))))
= **c**•**m**(M)
= m(M)***c**•**pV1**
I∧[M]∧(**c**|**m**;**dV1**) = lim (S[bn(M∩Pn)](**c**∧**m**(Pn(i,j,k))))
= **c**∧**m**(M)
= m(M)***c**∧**pV1**
I[M]*(**c**|**m**;**dV1**) = lim (S[bn(M∩Pn)](**c*****m**(Pn(i,j,k))))
= **c*****m**(M)
= m(M)***c*****pV1**.
In particular
I[M]*(1|**m**;**dR1**) = **m**(M).

Einstein, questioning: "Do sums and additions follow similar to derivatives?"

Breton: "Somewhat. For constant c1, **c1** and c2, **c2** with f1, **f1** and f2, **f2** bounded on M,

I[M]•(c1∗**f1**+c2∗**f2**|**m;dV1**)
= c1∗I[M]•(**f1**|**m;dV1**) + c2∗I[M]•(**f2**|**m;dV1**)
I[M]•(**c1**∗f1+**c2**∗f2|**m;dV1**)
= **c1**•I[M](f1|**m;dV1**) + **c2**•I[M]∗(f2|**m;dV1**)
I[M]•(**c1**∧**f1**+**c2**∧**f2**|**m;dV1**)
= **c1**•I[M]∧(**f1**|**m;dV1**) + **c2**•I[M]∧(**f2**|**m;dV1**)
I[M]•(**c1**∗**f1**+**c2**∗**f2**|**m;dV1**)
= **c1**∗I[M]•(**f1**|**m;dV1**) + **c2**∗I[M]•(**f2**|**m;dV1**)
I[M]∧(c1∗**f1**+c2∗**f2**|**m;dV1**)
= c1∗**I**[M]∧(**f1**|**m;dV1**) + c2∗**I**[M]∧(**f2**|**m;dV1**)
I[M]∧(**c1**∗f1+**c2**∗f2|**m;dV1**)
= **c1**∧**I**[M](f1|**m;dV1**) + **c2**∧**I**[M]∗(f2|**m;dV1**)
I[M]∧(**c1**∧**f1**+**c2**∧**f2**|**m;dV1**)
= **c1**•**T**(**I**[M]∗(**f1**|**m;dV1**)) – **c1**•**I**[M]∗(**f1**|**m;dV1**)
+ **c2**•**T**(**I**[M]∗(**f2**|**m;dV1**))– **c2**•**I**[M]∗(**f2**|**m;dV1**)
I[M]∗(c1∗f1+c2∗f2|**m;dV1**)
= c1∗**I**[M]∗(f1|**m;dV1**) + c2∗**I**[M]∗(f2|**m;dV1**)
I[M]∗(**c1**•**f1**+**c2**•**f2**|**m;dV1**)
= **c1**•**I**[M]∗(**f1**|**m;dV1**) + **c2**•**I**[M]∗(**f2**|**m;dV1**)
I[M]∗(c1∗**f1**+c2∗**f2**|**m;dV1**)
= c1∗**I[M]•(f1|m;dV1)** + **c2•I[M]•(f2|m;dV1)**
I[M]•(**c1**•**f1**+**c2**∗f2|**m;dV1**)
= **c1**∗**I**[M]∗(f1|**m;dV1**) + **c2**∗**I**[M]∗(f2|**m;dV1**)
I[M]∗(**c1**∧**f1**+**c2**∧**f2**|**m;dV1**)
= **C(c1)**•**I**[M]∗(**f1**|**m;dV1**)+ **C(c2)**•**I**[M]∗(**f2**|**m;dV1**).

Newton, admiringly: "We've learned how to write numerous equations quickly. But why include the measure in the notation?"

Breton: "Why indeed. Since the partition vector is specified in the notation, the reference to the measure in the symbol for integration is usually suppressed."

Local Integration of Local Derivatives

Einstein, leading: "How about integrals over gradients?"

Breton: "You ask a large question. Let's start simply. Just as continuous gradients may be expressed as sectional gradients (but not necessarily vice-versa) so may integrals of functions with continuous gradients be expressed in terms of a sectional partition.
For a subset of functions with basic derivatives and integrals, the following theorem applies."

Theorem (interior cancellation)

Given

f, a function with a basic, continuous gradient **D[v]*(f;dv)** in V3;

pV1 = p1***u1V1** + p2***u2V1** + p3***u3V1,** a partition vector;

I[**V3**]•(**f**;**dV1**) = g bounded;

for

dV1 = dV1***uV1**;

= d1V1***u1V1** + d2V1***u2V1** + d3V1***u3V1**;

= **pV1**/n^3;

UV1= **u1*u1V1** + **u2*u2V1** +**u3*u3V1**;

then

D[v]*(I[V3]•(f;dV1);dv) = **I[V3]•(D[v]*(f;dv);dV1)**

= **0**.

Proof:

Since g is constant, **D[v1]***(g;**dv**) = **0**.

Again, the bounded integral implies **f**(v***uv**)⟶**0** as v ⟶ ∞ for any **uv**.

Moreover,

I[V3]•(D[v1]*(f;dv);dV1)

= lim S[bn(V3∩Pn)](**pV1**•**T**[**D**[**v**(Pn(i,j,k))]*(**f**;**dv**)])/n^3

= lim S[bn(V3∩Pn)](**pV1**•[**UV1**$^{-1}$]•**T**[**UV1**]$^{-1}$
•**T**[**D**[**v**(Pn(i,j,k))]*(**f**;**dV1**)])/n^3

= lim S[bn(V3∩Pn)]
(d1V1*(**f**(**v**(Pn(i,j,k)+d1V1***u1V1**)) − **f**(**v**(Pn(i,j,k))))/d1V1
+ d2V1*(**f**(**v**(Pn(i,j,k)+d2V1***u2V1**))
− **f**(**v**(Pn(i,j,k))))/d2V1
+ d3V1*(**f**(**v**(Pn(i,j,k)+d3V1***u3V1**))
− **f**(**v**(Pn(i,j,k))))/d3V1)

= lim S[bn(V3∩Pn)](**f**(**v**(Pn(i+1,j,k))) − **f**(**v**(Pn(i,j,k)))
+ **f**(**v**(Pn(i,j+1,k))) − **f**(**v**(Pn(i,j,k)))
+ **f**(**v**(Pn(i,j,k+1))) − **f**(**v**(Pn(i,j,k))))

observing

v(Pn(i,j,k)+d1V1***u1V1**) = **v**(Pn(i,j,k)+p1V1***u1V1**/n^3)
= **v**(Pn(i+1,j,k)).

The sum over the index i, for illustration, becomes

(... + **f**(**v**(Pn(i+2,j,k))) − **f**(**v**(Pn(i+1,j,k)))
+ **f**(**v**(Pn(i+1,j,k))) − **f**(**v**(Pn(i,j,k)))
+ **f**(**v**(Pn(i,j,k)))
− **f**(**v**(Pn(i−1,j,k)))
+...)

⟶ (**f**(v***u1V1**) + **f**(v*(−**u1V1**))) as v ⟶ ∞

= **0**.

The same is true for indices j and k.

The same is true for any **uv** = **u**(**v**(Pn(i,j,k))) since **uv** equals some weighted combination of the indices.

Therefore, **I[V3]•(D[v1]*(f;dv);dV1)** = **0**.

qed

The rule n^3***dV1** = **pV1** is called the **matched integration rule**."

Newton: "This result easily extends to sectional gradients."

Breton: "Yes, We can state the result as a corollary."

Corollary
Given:
f(v), a function with a basic sectional gradient everywhere in **V3** with bounded invergence;
then

$$\mathbf{I}[\mathbf{V3}]\bullet(\mathbf{D}[\mathbf{v}]*(\mathbf{f};\mathbf{dV1})\bullet[\mathbf{UV1}]^{-1}\bullet\mathbf{T}[\mathbf{UV1}]^{-1};\mathbf{dV1}) = \mathbf{0}$$

Einstein: "What about special cases?."

Breton: "Suppose for a given partition based on **pV1**, a gradient **D[v]*(f;dV2)** is constant over M. Then

$$\begin{aligned}
&I[M]\bullet(\mathbf{D}[\mathbf{v}]*(\mathbf{f};\mathbf{dV2});\mathbf{dV1})\\
&\quad= \lim S[bn(M\cap Pn)](\mathbf{m}(Pn(i,j,k))\bullet\mathbf{T}[\mathbf{D}[\mathbf{v}(Pn(i,j,k))]*(\mathbf{f};\mathbf{dV2})])\\
&\quad= \lim S[bn(M\cap Pn)](\mathbf{m}(Pn(i,j,k)))\bullet\mathbf{T}[\mathbf{D}[\mathbf{v}]*(\mathbf{f};\mathbf{dV2})]\\
&\quad= \mathbf{m}(M|\mathbf{pV1})\bullet\mathbf{T}[\mathbf{D}[\mathbf{v}]*(\mathbf{f};\mathbf{dV2})]\\
&\quad= vol(M)*\mathbf{pV1}\bullet\mathbf{T}[\mathbf{D}[\mathbf{v}]*(\mathbf{f};\mathbf{dV2})]/vol(P1(i,j,k)).
\end{aligned}$$

Newton, following closely: "And this result can be stated for simply continuous functions."

Breton: "Yes. Moreover, for **f(v)** a function with a basic, simply continuous *constant* gradient

$$\begin{aligned}
&I[M]\bullet(\mathbf{D}[\mathbf{v}]*(\mathbf{f};\mathbf{dV2})\bullet[\mathbf{UV2}]^{-1}\bullet\mathbf{T}[\mathbf{UV2}]^{-1};\mathbf{dV1})\\
&\quad= vol(M)*\mathbf{pV1}\bullet[\mathbf{UV2}]^{-1}\bullet\mathbf{T}[\mathbf{UV2}]^{-1}\bullet\mathbf{T}[\mathbf{D}[\mathbf{v}]*(\mathbf{f};\mathbf{dV2})]\\
&\qquad\qquad /vol(P1(i,j,k))\\
&\quad= vol(M)*\mathbf{pV1}\bullet\mathbf{T}[\mathbf{D}[\mathbf{v}]*(\mathbf{f};\mathbf{dv})]/vol(P1(i,j,k)).
\end{aligned}$$

Einstein: "And if the function is not constant?"

Breton: "Now you are asking a larger question with a surprising result. More generally, for **f** continuously differentiable

$$\begin{aligned}
&I[M]\bullet(\mathbf{D}[\mathbf{v}]*(\mathbf{f};\mathbf{dV1})\bullet[\mathbf{UV1}]^{-1}\bullet\mathbf{T}[\mathbf{UV1}]^{-1};\mathbf{dV1})\\
&\quad= \lim S[bn(M\cap Pn)](\mathbf{m}(Pn(i,j,k))\\
&\qquad\quad \bullet[\mathbf{UV1}]^{-1}\bullet\mathbf{T}[\mathbf{UV1}]^{-1}\bullet\mathbf{T}[\mathbf{D}[\mathbf{v}(Pn(i,j,k))]*(\mathbf{f};\mathbf{dV1})])
\end{aligned}$$

= lim S[bn(M∩Pn)]((p1V1***u1** + p2V1***u2** + p3V1***u3**)
/n^3)
•(**u1***(**f**(**v**(Pn(i,j,k))+d1V1***u1V1**)
- **f**(**v**(Pn(i,j,k))))/d1V1
+ **u2***(**f**(**v**(Pn(i,j,k))+d2V1***u2V1**)
- **f**(**v**(Pn(i,j,k))))/d2V1
+ **u3***(**f**(**v**(Pn(i,j,k))+d3V1***u3V1**)
-**f**(**v**(Pn(i,j,k))))/d3V1)
= lim S[bn(M∩Pn)](**f**(**v**(Pn(i,j,k))+d1V1***u1V1**)
- **f**(**v**(Pn(i,j,k)))
+ **f**(**v**(Pn(i,j,k))+d2V1***u2V1**) - **f**(**v**(Pn(i,j,k)))
+ **f**(**v**(Pn(i,j,k))+d3V1***u3V1**) - **f**(**v**(Pn(i,j,k))))
= lim S[bn(M∩Pn)](**f**(**v**(Pn(i,j,k)+**dV1**)) - **f**(**v**(Pn(i,j,k))))
= lim (S[bn(SMPE|**pV1**∩Pn)](**f**(**v**(Pn(i,j,k))+**dV1**))
- S[bn(SMPI|**pV1**∩Pn)](**f**(**v**(Pn(i,j,k))))
≡ I[SMPI|**pV1**,SMPE|**pV1**](**f**(**sv**)).

The above equation is the **fundamental theorem of vector calculus for sectional gradients over a set M in V3**."

Einstein: "What happens if the function is further restricted?"

Breton: "If further **f** is asymptotic
I[**V3**]•(**D**[**v**]*(**f**;**dV1**)•[**UV1**]$^{-1}$•**T**[**UV1**]$^{-1}$;**dV1**)
= I[M∪SM]•(**D**[**v**]*(**f**;**dV1**)•[**UV1**]$^{-1}$•**T**[**UV1**]$^{-1}$;**dV1**)
+ I[**V3**-(M∪SM)]•(**D**[**v**]*(**f**;**dV1**)•[**UV1**]$^{-1}$•**T**[**UV1**]$^{-1}$;**dV1**)
= **0**.
Thus for this case
I[**V3**-(M∪SM)]•(**D**[**v**]*(**f**;**dV1**)•[**UV1**]$^{-1}$•**T**[**UV1**]$^{-1}$;**dV1**)
= -I[SMPI|**pV1**,SMPE|**pV1**](**f**(**sv**)).

Newton, adding the usual: "Results for divergences and curls must follow the same path."

Breton: "For functions for which cancellation operates as above, sectional divergences and curls are expressed similarly.
I[M]•(**D**[**v**]∧(**f**;**dV1**);**dV1**)
= I[SMPI|**pV1**,SMPE|**pV1**](**D**[**sv**]∧(**f**;**dV1**)•**dV1**)
I[M]∧(**D**[**v**]∧(**f**;**dV1**);**dV1**)
= I[SMPI|**pV1**,SMPE|**pV1**](**D**[**sv**]∧(**f**;**dV1**)∧**dV1**)
I[M]*(**D**[**v**]•(**f**;**dV1**);**dV1**)
= I[SMPI|**pV1**,SMPE|**pV1**](**D**[**sv**]•(**f**;**dV1**)***dV1**)
I[M]*(**D**[**v**]∧(**f**;**dV1**);**dV1**)
= I[SMPI|**pV1**,SMPE|**pV1**](**D**[**sv**]∧(**f**;**dV1**)***dV1**)
and for scalar functions
I[M]•(**D**[**v**]*(f;**dV1**)•[**UV1**]$^{-1}$•**T**[**UV1**]$^{-1}$;**dV1**)
= I[SMPI|**pV1**,SMPE|**pV1**](f(**sv**))

I[M]∧(**D**[**v**]∗(f;**dV1**);**dV1**)
= I[SMPI|**pV1**,SMPE|**pV1**](**D**[**sv**]∗(f;**dV1**)∧**dV1**)
I[M]∗(**D**[**v**]∗(f;**dV1**);**dV1**)
= I[SMPI|**pV1**,SMPE|**pV1**]((**D**[**sv**]∗(f;**dV1**)∗**dV1**).

These equations are extensions of the **fundamental theorem of the calculus for V3**."

Einstein, pleased with his ability to direct the conversation: "Other analogies?"

Breton: "Now let {Mi} be a properly nested sequence of subsets of M.
For **f**(**v**) be continuous over M, let

G(Mi) ≡ **I**[Mi]∗(**f**;**dV1**).

Then for Mi ⊂ Mj
I[Mj − Mi]∗(**f**;**dR1**) = **I**[Mj]∗(**f**;**dR1**) − **I**[Mi]∗(**f**;**dR1**)
= **G**(Mj) − **G**(Mi).

Einstein, looking for another direction: "Can you use step functions?"

STEP FUNCTIONS IN V3

Breton: "In **V3** the idea of order is incompatible with the usual topology. As substitute, consider the interior, exterior, and surface cells of a given measurable set M in a given partition. Then **V3** can be separated into parts over whose surface SM the discontinuity in step functions may be defined."

Einstein: "Be specific."

Breton: "It is useful to name the following sets

EMPEF|**pV1** ≡ {**v**|**v**=**ves**+dV1∗**uV1**, **ves** in SMPE|**pV1**; dV1>0}
EMPEB|**pV1** ≡ {**v**|**v**=**ves**+dV1∗**uV1**, **ves** in SMPE|**pV1**; dV1<0}
EMPIF|**pV1** ≡ {**v**|**v**=**vis**+dV1∗**uV1**, **vis** in SMPI|**pV1**; dV1>0}
EMPIB|**pV1** ≡ {**v**|**v**=**vis**+dV1∗**uV1**, **vis** in SMPI|**pV1**; dV1<0}
EMNEF|**pV1** ≡ {**v**|**v**=**ves**+dV1∗**uV1**, **ves** in SMNE|**pV1**; dV1>0}
EMNEB|**pV1** ≡ {**v**|**v**=**ves**+dV1∗**uV1**, **ves** in SMNE|**pV1**; dV1<0}
EMNIF|**pV1** ≡ {**v**|**v**=**vis**+dV1∗**uV1**, **vis** in SMNI|**pV1**; dV1>0}
EMNIB|**pV1** ≡ {**v**|**v**=**vis**+dV1∗**uV1**, **vis** in SMNI|**pV1**; dV1<0}.

Einstein: "An illustration would help."

Breton: "Let me illustrate just EMPEF|**pV1**."

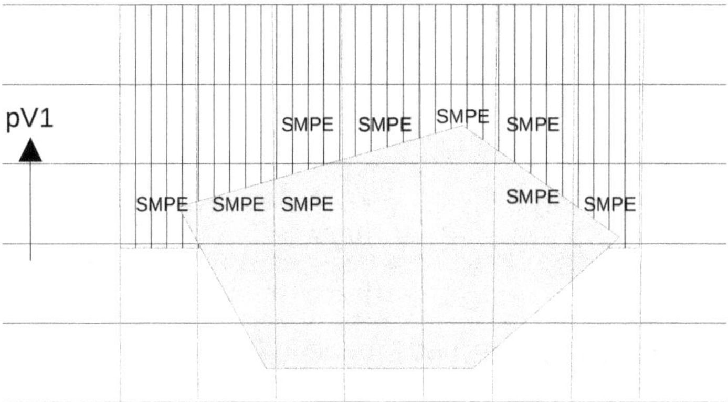

The illustration shows all the SMPE cells. The hatch area illustrates EMPEF|**pV1**, the vicinity of all vectors in these cells extended in the **pV1** direction."

Newton: "The set EMPEB|**pV1** are just the same SMPE cells extended in the negative direction."

Einstein: "Together they include all vectors in V3."

Breton: "Something similar to the step function u(q). The function has a constant value for every q, one constant for the forward direction, and second constant for the negative direction."

Newton: "So by defining EMPEF|**pV1** and EMPEB|**pV1** in forward and backward directions relative to **PV1** you have constructed a substrate on which to define step functions locally in V3."

Einstein: "So on with some definitions."

Breton: "Expect more than one."

Point Step Functions in V3

Definition (point step functions in V3)

Given

pV1 ≡ pV1***uV1,** a partition vector;
Pn, a set of nested partitions based on **pV1**;
M, a measurable set;
SM, the surface of M;

then

for a given **sev** in SMPE|**pV1**

$$
\begin{aligned}
us(\mathbf{v}-\mathbf{sev}|M) &\equiv 0, & \mathbf{v} = \mathbf{sev}+dV1*\mathbf{uV1}, \; dV1 \le 0\\
&\equiv 1, & \text{otherwise}\\
vs(\mathbf{v}-\mathbf{sev}|M) &\equiv 0, & \mathbf{v} = \mathbf{sev}+dV1*\mathbf{uV1}, \; dV1 < 0\\
&\equiv 1, & \text{otherwise}
\end{aligned}
$$

for a given **siv** in SMPI|**pV1**

$$
\begin{aligned}
us(\mathbf{v}-\mathbf{siv}|M) &\equiv 0, & \mathbf{v} = \mathbf{siv}+dV1*\mathbf{uV1}, \; dV1 \le 0\\
&\equiv 1, & \text{otherwise}\\
vs(\mathbf{v}-\mathbf{siv}|M) &\equiv 0, & \mathbf{v} = \mathbf{siv}+dV1*\mathbf{uV1}, \; dV1 < 0\\
&\equiv 1, & \text{otherwise}
\end{aligned}
$$

are **unit point step functions in V3**.

end of definition

Einstein: "Would you favor us with an illustration?"

Breton: "I'll try. Again the illustration uses the same two dimensional figure to represent a three dimension vicinity; Further, the illustration shows only the us(**v**–**sev**|M) values. These are shown as a partition of a plane, whereas they are partitions of **V3**."

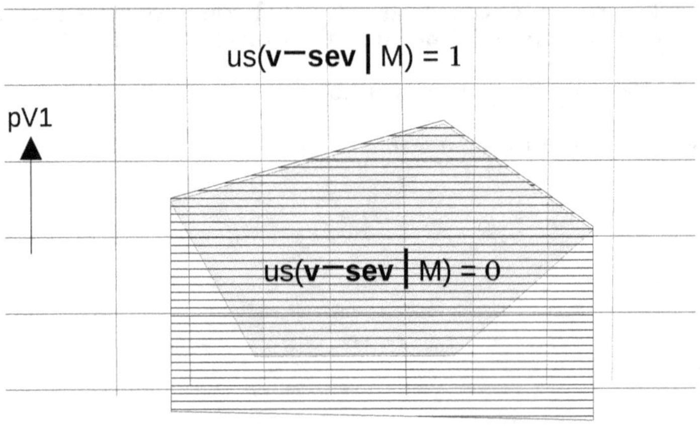

Newton: "The other unit point step functions relate to different partitions."

Breton: "Correct."

Einstein: "And how do you use these step functions."

Breton: "First let us calculate the directional gradient of these step functions in the various parts of the partition. Let

$$\mathbf{dV1} \equiv dV1*\mathbf{uV1}$$

For a given **sev** in SMPE | **pV1**

$$\mathbf{D}[\mathbf{v},\mathbf{v}+\mathbf{dV1}]*(us(\mathbf{v}-\mathbf{sev})|M;\mathbf{dV1}) = \mathbf{0} \qquad \mathbf{v} \neq \mathbf{sev}$$
$$\mathbf{D}[\mathbf{v},\mathbf{v}-\mathbf{dV1}]*(us(\mathbf{v}-\mathbf{sev})|M;\mathbf{dV1}) = \mathbf{0} \qquad \mathbf{v} \neq \mathbf{sev}$$
$$\mathbf{D}[\mathbf{v},\mathbf{v}+\mathbf{dV1}]*(us(\mathbf{v}-\mathbf{sev})|M;\mathbf{dV1}) = \lim \mathbf{uV1}/dV1 \qquad \mathbf{v} = \mathbf{sev}$$
$$\mathbf{D}[\mathbf{v},\mathbf{v}-\mathbf{dV1}]*(us(\mathbf{v}-\mathbf{sev})|M;\mathbf{dV1}) = \mathbf{0} \qquad \mathbf{v} = \mathbf{sev}.$$

Similarly,

$$\mathbf{D}[\mathbf{v},\mathbf{v}+\mathbf{dV1}]*(vs(\mathbf{v}-\mathbf{sev})|M;\mathbf{dV1}) = \mathbf{0} \qquad \mathbf{v} \neq \mathbf{sev}$$
$$\mathbf{D}[\mathbf{v},\mathbf{v}-\mathbf{dV1}]*(vs(\mathbf{v}-\mathbf{sev})|M;\mathbf{dV1}) = \mathbf{0} \qquad \mathbf{v} \neq \mathbf{sev}$$
$$\mathbf{D}[\mathbf{v},\mathbf{v}+\mathbf{dV1}]*(vs(\mathbf{v}-\mathbf{sev})|M;\mathbf{dV1}) = \mathbf{0} \qquad \mathbf{v} = \mathbf{sev}$$
$$\mathbf{D}[\mathbf{v},\mathbf{v}-\mathbf{dV1}]*(vs(\mathbf{v}-\mathbf{sev})|M;\mathbf{dV1}) = -\lim \mathbf{uV1}/dV1 \qquad \mathbf{v} = \mathbf{sev}.$$

A similar result holds for *any* **sev** in SMPE | **pV1**."

Newton, elaborating: "We're building a structure similar to the usual step functions."

Breton: "With extensions that look like three dimensional impulse functions."

Einstein, impatiently: "How do they work with functions?"

Breton: "Let **f** be a generalized function over **V3**. Step functions can be multiplied with the function in different ways. Furthermore results can be obtained for various parts of a partition.

For **sev** in SMPE | **pV1**

$$\mathbf{D}[\mathbf{v},\mathbf{v}+\mathbf{dV1}]*(\mathbf{f}(\mathbf{v})*us(\mathbf{v}-\mathbf{sev})|M;\mathbf{dV1})$$
$$= \mathbf{D}[\mathbf{v},\mathbf{v}+\mathbf{dV1}]*(\mathbf{f};\mathbf{dV1}) \qquad \mathbf{v} \neq \mathbf{sev}+dV1*\mathbf{uV1}$$
$$= \mathbf{D}[\mathbf{v},\mathbf{v}+\mathbf{dV1}]*(\mathbf{f};\mathbf{dV1}) \qquad \mathbf{v} = \mathbf{sev}+dV1*\mathbf{uV1} \quad dV1 > 0$$
$$= [0] \qquad \mathbf{v} = \mathbf{sev}+dV1*\mathbf{uV1} \quad dV1 < 0$$
$$= \lim \mathbf{f}(\mathbf{sev}+\mathbf{dV1})*\mathbf{uV1}/dV1 \qquad \mathbf{v} = \mathbf{sev}$$

$$\mathbf{D}[\mathbf{v},\mathbf{v}-\mathbf{dV1}]*(\mathbf{f}(\mathbf{v})*us(\mathbf{v}-\mathbf{sev})|M;\mathbf{dV1})$$
$$= \mathbf{D}[\mathbf{v},\mathbf{v}-\mathbf{dV1}]*(\mathbf{f};\mathbf{dV1}) \qquad \mathbf{v} \neq \mathbf{sev}+dV1*\mathbf{uV1}$$
$$= \mathbf{D}[\mathbf{v},\mathbf{v}-\mathbf{dV1}]*(\mathbf{f};\mathbf{dV1}) \qquad \mathbf{v} = \mathbf{sev}+dV1*\mathbf{uV1} \quad dV1 > 0$$
$$= [0] \qquad \mathbf{v} = \mathbf{sev}+dV1*\mathbf{uV1} \quad dV1 \leq 0$$

$$\mathbf{f}(\mathbf{v})*\mathbf{D}[\mathbf{v},\mathbf{v}+\mathbf{dV1}]*(us(\mathbf{v}-\mathbf{sev})|M;\mathbf{dV1})$$
$$= [0] \qquad \mathbf{v} \neq \mathbf{sev}$$
$$= \lim \mathbf{f}(\mathbf{sev})*\mathbf{uV1}/dV1 \qquad \mathbf{v} = \mathbf{sev}$$

f(v)*D[v,v-dV1]*(us(v-sev)|M;dV1)
= [0]
(us(v-sev)|M)*D[v,v+dV1]*(f;dV1)
= D[v,v+dV1]*(f;dV1)   v ≠ sev+dV1*uV1
= D[v,v+dV1]*(f;dV1)   v = sev+dV1*uV1 dV1>0
= [0]   v = sev+dV1*uV1 dV1≤ 0
(us(v-sev)|M)*D[v,v-dV1]*(f;dV1)
= D[v,v-dV1]*(f;dV1),   v ≠ sev+dV1*uV1
= D[v,v-dV1]*(f;dV1),   v = sev+dV1*uV1, dV1 > 0
= [0],   v = sev+dV1 *uV1, dV1≤ 0.
Likewise for vs(v-sev)|M
D[v,v+dV1]*(f(v)*vs(v-sev)|M;dV1)
= D[v,v+dV1]*(f;dV1)   v ≠ sev+dV1*uV1
= D[v,v+dV1]*(f;dV1)   v = sev+dV1*uV1 dV1 ≥ 0
= 0,   v = sev+dV1*uV1 dV1 < 0
D[v,v-dV1]*(f(v)*vs(v-sev)|M;dV1)
= D[v,v-dV1]*(f;dV1)   v ≠ sev+dV1*uV1
= D[v,v-dV1]*(f;dV1)   v = sev+dV1*uV1 dV1 > 0
= [0]   v = sev+dV1*uV1 dV1 < 0
= lim -f(sev)*uV1/dV1   v = sev
f(v)*D[v,v+dV1]*(vs(v-sev)|M;dV1)
= [0]
f(v)*D[v,v-dV1]*(vs(v-sev)|M;dV1)
= [0]   v ≠ sev
= lim -f(sev)*uV1/dV1   v = sev
(vs(v-sev)|M)*D[v,v+dV1]*(f;dV1)
= D[v,v+dV1]*(f;dV1)   v ≠ sev+dV1*uV1
= D[v,v+dV1]*(f;dV1)   v = sev+dV1*uV1 dV1 ≥ 0
= [0]   v = sev+dV1*uV1 dV1 < 0
(vs(v-sev)|M)*D[v,v-dV1]*(f;dV1)
= D[v,v-dV1]*(f;dV1)   v ≠ sev+dV1*uV1
= D[v,v-dV1]*(f;dV1)   v = sev+dV1*uV1 dV1 ≥ 0
= [0]   v = sev+dV1*uV1 dV1 < 0.

Similar results hold for **siv** in SMPI|**pV1**."

Newton, reflecting: "These results mimic previous results."

Breton: "Higher up the mountain we find different but parallel trails."

Einstein: "A nice metaphor. Do we continue with more restricted functions?"

Breton: "You know the trail.

For **f** continuous at **sev**
D[v,v+dV1]*(f(v)*us(v-sev)|M;dV1)
= f(v)*D[v,v+dV1]*(u(v-sev)|M;dV1)
+ vs(v-sev)|M*D[v,v+dV1]*(f;dV1)
D[v,v+dV1]*(f(v)*vs(v-sev)|M;dV1)
= f(v)*D[v,v+dV1]*(vs(v-sev)|M;dV1)
+ vs(v-sev)|M*D[v,v+dV1]*(f;dV1)
D[v,v-dV1]*(f(v)*us(v-sev)|M;dV1)
= f(v)*D[v,v-dV1]*(us(v-sev)|M;dV1)
+ us(v-sev)|M*D[v,v-dV1]*(f;dV1)
D[v,v-dV1]*(f(v)*vs(v-sev)|M;dV1)
= f(v)*D[v,v-dV1]*(vs(v-sev)|M;dV1)
+ vs(v-sev)|M*D[v,v-dV1]*(f;dV1).
Corresponding statements may be made for ***sir*** in SMPI."

Einstein: "The path opens up the possibility of integration."

Breton: "You know the trail well indeed."
Now let **f** be a generalized function over **V3** and let
pV1 ≡ pV1*uV1
be a partition vector with corresponding set of partitions Pn.
Then
lim(S[bn(V3∩Pn)](f(v(Pn(i,j,k)))
D[v,v+dV1](us(v-sev)|M;dV1)•m(Pn(i,j,k))
= lim lim f(sev)*pV1/(n^3*dV1), n^3*dV1 = pV1.
With the limits so linked,
I[V3]•(f(v)*D[v,v+dV1]*(us(v-sev)|M;dV1);dV1)
= f(sev) sev in SMPE|pV1.
Likewise,
I[V3]•(f(v)*D[v,v+dV1]*(us(v-siv)|M;dV1);dV1)
= f(siv) siv in SMPI|pV1
I[V3]•(f(v)*D[v,v-dV1]*(vs(v-sev)|M;dV1);dV1)
= -f(sev) sev in SMPE|pV1
I[V3]•(f(v)*D[v,v-dV1]*(vs(v-siv)|M;dV1);dV1)
= -f(siv) siv in SMPI|pV1.
These equations may be taken as functions over subsets of SM."

Einstein: "Now let **f** be continuous over V3."

Breton: "Then
I[V3]•(us(v-sev)*D[v,v+dV1]*(f(v);dV1)|M;dV1)
= lim S[bn(V3∩Pn)]((us(v(Pn(i,j,k))-sev)|M)
D[v(Pn(i,j,k)),v(Pn(i,j,k));dV1](f(v);dV1)•m(Pn(i,j,k)))
= lim lim -f(sev+dV1)*pV1/(n^3*dV1), n^3*dV1 = pV1
= -f(sev).
Likewise,
I[V3]•(us(v-siv)*D[v,v+dV1]*(f(v)|M;dV1);dV1)
= -f(siv)

I[V3]•(vs(v−sev)∗D[v,v−dV1]∗(f(v)|M;dV1);dV1)
= f(sev)
I[V3]•(vs(v−siv)∗D[v,v−dV1]∗(f(v)|M;dV1);dV1)
= f(siv).
Further still for I[V3]•(f;dV1) bounded,
I[V3]•(D[v,v+dV1]∗(((us(v−sev)|M)∗f);dV1);dV1)
= lim lim S[bn(V3∩Pn)]
(f(v(Pn(i+1,j+1,k+1)))
∗(us(v(Pn(i+1,j+1,k+1))−sev)|M)
−f(v(Pn(i,j,k)))
∗(us(v(Pn(i,j,k))−sev)|M))
= f(v(Pn(i1+1,j1+1,k1+1)))
∗(us(v(Pn(i1+1,j1+1,k1+1))−sev)|M)
−f(v(Pn(i1,j1,k1)))
∗(us(v(Pn(i1,j1,k1))−sev)|M)
+ f(v(Pn(i1+2,j1+2,k1+2)))
∗(us(v(Pn(i1+2,j1+2,k1+2))−sev)|M)
−f(v(Pn(i1+1,j1+1,k1+1)))
∗(us(v(Pn(i1+1,j1+1,k1+1))−sev)|M)
+ ...
= 0.

Likewise,
I[V3]•(D[v,v+dV1]∗((us(v−siv)|M)∗f;dV1);dV1) = 0
I[V3]•(D[v,v−dV1]∗((vs(v−sev)|M)∗f;dV1);dV1) = 0
I[V3]•(D[v,v−dV1]∗((vs(v−siv)|M)∗f;dV1);dV1) = 0.

Einstein: "Other integral operations are possible."

Breton: "Right on.
I[V3]•(f(v)∧D[v,v+dV1]∗(us(v−sev)|M;dV1);dV1)
= lim (f(sev)∧D[v]∗(us(v−sev)|M);dV1)•m(Pn(i,j,k))
= f(sev)•(D[v,v+dV1]∗(us(v−sev)|M;dV1)∧m(Pn(i,j,k)))
= 0
I[V3]∧(f(v)∗D[v,v+dV1]∗(us(v−sev)|M;dV1);dV1)
= 0
I[V3]∧(f(v)∧D[v,v+dV1]∗(us(v−sev)|M;dV1);dV1)
= f(sev)•((uV1∗uV1) − I)
I[V3]∗(f(v)∗D[v,v+dV1]∗(us(v−sev)|M;dV1);dV1)
= f(sev)∗(uV1∗uV1)
I[V3]∗(f(v)∧D[v,v+dV1]∗(us(v−sev)|M;dV1);dV1)
= C(f(sev))•(uV1)∗(uV1)
I[V3]∗(f(v)•D[v,v+dV1]∗(us(v−sev)|M;dV1);dV1)
= f(sev)•(uV1∗uV1).

For functions for which cancellation holds, non-vector integration gives similar results. Similar results hold for vs(v−sev) and siv."

Newton: "Our results so far have to do with edges of M. Why not step functions over the whole of M?"

Local Step Functions in V3

Breton: "Yes, there is a path that way too. We call these step functions 'local' to distinguish them from 'point'."

Definition (local step functions in V3)

Given

pV1 ≡ pV1∗**uV1,** a partition vector;
Pn, a set of nested partitions based on **pV1**;
M, a measurable set;
SM, the surface of M;

then

$$\begin{aligned} us(\mathbf{v})|M &\equiv 0, \quad \mathbf{v} \text{ in } M \cup SM \\ &\equiv 1, \quad \text{otherwise} \\ vs(\mathbf{v})|M &\equiv 0, \quad \mathbf{v} \text{ in } M - SM \\ &\equiv 1, \quad \text{otherwise} \end{aligned}$$

are **local unit step functions in V3**

end of definition

Einstein: "How about an illustration?"

Breton: "With the usual restrictions."

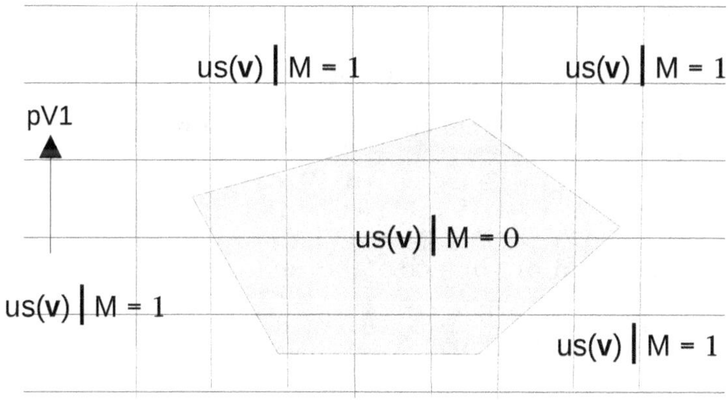

The local step function of every vector in **V3** has a value of 1, except for vectors in M where the step function equals 0. The 'us' step function keeps the zero for the boundary of M; the 'vs' step function does not."

Newton: "This is an elegant extension of the first definition of step functions."

Einstein: "So where do we go from here?"

Breton: "First let's calculate the gradients.
For **dV1** ≡ d1V1***u1V1** + d2V1***u2V1** + d3V1***u3V1**
Then
D[v]*(us(**v**|M);**dV1**) = **0** **v** in M−SM
D[v]*(us(**v**|M);**dV1**) = **0** **v** in V3−(M∪SM)
D[v]*(us(**v**|M);**dV1**) = lim(us(**v**+d1V1***u1V1**|M)***u1V1**/d1V1
+ us(**v**+d2V1***u2V1**|M)***u2V1**/d2V1
+ us(**v**+d3V1***u3V1**|M)***u3V1**/d3V1),
v in SM.
For **v** in SM then, **D[v]***(us(**v**|M);**dV1**) may equal **0**, or may be impulsive in one, two, or all three sectional directions.
Notice the difference for vs.
D[v]*(vs(**v**|M);**dV1**) = **0** **v** in M−SM
D[v]*(vs(**v**|M);**dV1**) = **0** **v** in V3−(M∪SM)
D[v]*(vs(**v**|M);**dV1**) = lim ((vs(**v**+d1V1***u1V1**|M)−1)***u1V1**)
/d1V1
+ ((v(**v**+d2V1***u2V1**|M)−1)***u2V1**)/d2V1
+ ((v(**v**+d3V1***u3V1**|M)−1)***u3V1**)/d3V1,
v in SM.

Einstein: "What is the difference?"

Breton: "Look closely. Both step functions have identical values inside M and outside M. The only difference occurs on the boundary. On the boundary us(**v**|M equals 0 while vs(**v**|M equals 1."

Newton: "Just as we should expect for step functions."

Breton: "Now check the gradients. For us(**v**|M, one branch of the limit would be more completely written as
(us(**v**+d1V1***u1V1**|M) − us(sv))***u1V1**/d1V1
= (us(**v**+d1V1***u1V1**|M) − 0)***u1V1**/d1V1
= us(**v**+d1V1***u1V1**|M)***u1V1**/d1V1
whereas for vs(**v**|M, one branch of the limit would be more completely written as
(vs(**v**+d1V1***u1V1**|M) − vs(**sv**))***u1V1**/d1V1
= (vs(**v**+d1V1***u1V1**|M) −1)***u1V1**/d1V1.

Newton: "So what looks like an arbitrary change, actually is just an application of the definitions."

Breton: "For **v** in SM then, **D[v]***(vs(**v**|M);**dV1**) may equal **0**, or may be impulsive in one, two, or all three sectional directions."

Einstein: "Are there no differences?"

Breton: "Yes there are. Let **sv** be an element of SM and **v** = **sv**+diV1***uiV1**. The following table gives the relationship between the differences in us(**v**|M) and vs(**v**|M).

Location of **v**	us(**v**|M)-us(**sv**|M)	vs(**v**|M)-vs(**sv**|M)
V3-(M∪SM)	1	0
SM	0	0
M-SM	0	-1

Local Step Functions at Surfaces

Thus where the gradient of us(**sv**|M) is impulsive, it is always positively impulsive, whereas where the gradient of vs(**sv**|M) is impulsive, it is always negatively impulsive. Moreover in those directions where the gradient of us(**sv**|M) is impulsive, the gradient of vs(**sv**|M) is zero, and vice-versa.

Clearly, D[v]*(us(v|M);dV1) and D[v]*(vs(v|M);dV1) are *not* continuously differentiable."

Einstein, genuinely intrigued: "And how does this affect products of step functions with other functions?"

Breton: "Again there are many combinations. Now let **f** be a generalized function over **V3**. For the various gradients and integrals of such gradients we have.

D[v]*(f(v)*us(v)|M;dV1)

= D[v]*(f;dV1) v in V3-(M∪SM)

= lim f(v+d1V1*u1V1)*us(v+d1V1*u1V1)*u1V1/d1V1

+ f(v+d2V1*u2V1)*us(v+d2V1*u2V1)*u2V1/d2V1

+ f(v+d3V1*u3V1)*us(v+d3V1*u3V1)*u3V1/d3V1

v in SM

= [0] v in M-SM

f(v)*D[v]*(us(v)|M;dV1)

= [0] v not in SM

= f(v)* (lim us(v+d1V1*u1V1)*u1V1/d1V1

+ us(v+d2V1*u2V1)*u2V1/d2V1

+ us(v+d3V1*u3V1)*u3V1/d3V1)

v in SM

(us(v)|M)*D[v]*(f;dV1)

= D[v]*(f;dV1) v in V3-(M∪SM)

= [0] v in SM

= [0] v in M-SM

D[v]*(f(v)*vs(v)|M;dV1)
= D[v]*(f;dV1) v in V3-(M∪SM)
= lim (f(v+d1V1*u1V1)*v(sv+d1V1*u1V1)*u1V1-f(v))
/d1V1
+(f(v+d2V1*u2V1)*vs(v+d2V1*u2V1)*u2V1-f(v))
/d2V1
+(f(v+d3V1*u3V1)*vs(v+d3V1*u3V1)*u3V1-f(v))
/d3V1
v in SM
= [0] v in M–SM
f(v)*D[v]*(vs(v)|M;dV1)
= [0] v not in SM
= f(v)* (lim (vs(v+d1V1*u1V1)-1)*u1V1/d1V1
+ (vs(v+d2V1*u2V1)-1)*u2V1/d2V1
+(vs(v+d3V1*u3V1)-1)*u3V1/d3V1)
v in SM
(vs(v)|M)*D[v]*(f;dV1)
= D[v]*(f;dV1) v in V3-(M∪SM)
= D[v]*(f;dV1) v in SM
= [0] v in M–SM.

Einstein: "And for integration?"

Breton: "Now let
pV1 ≡ p1V1*u1V1 + p21V1*u2V1 +p3V1*u3V1
be a partition vector with corresponding set of partitions Pn and
dV1 ≡ u1V1/d1V1 + u2V1/d2V1 +u3V1/d3V1
p1 ≡ p1V1*u1 + p2V1*u2 +p3V1*u3
d1 ≡ u1/d1V1 + u2/d1V1 +u3/d1V1.
Then
lim S[bn(V3∩Pn)](f(v(Pn(i,j,k)))
(D[v(Pn(i,j,k)))](us(v)|M;dV1)
+ D[v(Pn(i,j,k))]*(v(v)|M;dV1))
•[UV1]$^{-1}$•T[UV1]$^{-1}$
•m(Pn(i,j,k))
= lim S[bn(SM∩Pn)](f(sv(Pn(i,j,k)))
(D[sv(Pn(i,j,k))](us(v)|M;dV1)
+ D[sv(Pn(i,j,k))]*(vs(v)|M;dV1))
•m(Pn(i,j,k)))
= lim S[bn(SM∩Pn)](f(sv(Pn(i,j,k)))
*(((us(sv+d1V1*u1V1)|M
+ vs(sv+d1V1*u1V1)|M-1)*u1V1/d1V1)
+ ((us(sv+d2V1*u2V1)|M
+ vs(sv+d2V1*u2V1)|M-1)*u2V1/d2V1)
+ ((us(sv+d3V1*u3V1)|M
+ vs(sv+d3V1*u3V1)|M-1)*u3V1/d3V1))
•[UV1]$^{-1}$•T[UV1]$^{-1}$
•(p1V1*u1V1 + p2V1*u2V1 +p3V1*u3V1)/n^3

= lim S[bn(SM∩Pn)](**f**(**sv**(Pn(i,j,k)))
*((us(**sv**+d1V1***u1V1**)|M
+ vs(**sv**+d1V1***u1V1**)|M-1)
*(p1V1/d1V1)/n^3
+ (us(**sv**+d2V1***u2V1**)|M
+ vs(**sv**+d2V1***u2V1**)|M-1)
*(p2V1/d2V1)/n^3
+ (us(**sv**+d3V1***u3V1**)|M
+ vs(**sv**+d3V1***u3V1**)|M-1)
*(p3V1/d3V1)/n^3)).

Let the limits be taken with

diV1 = piV1/n^3.

With the limits so linked,

I[**V3**]•(**f**(**v**)*(**D**[**v**]*(us(**v**)|M;**dV1**) + **D**[**v**]*(vs(**v**)|M;**dV1**))
•[**UV1**]$^{-1}$•**T**[**UV1**]$^{-1}$;**dV1**)
= lim S[bn(SM∩Pn)](**f**(**sv**(Pn(i,j,k)))
*(us(**sv**+d1V1***u1V1**)|M
+ vs(**sv**+d1V1***u1V1**)|M-1
+ us(**sv**+d2V1***u2V1**)|M
+ vs(**sv**+d2V1***u2V1**)|M-1
+ us(**sv**+d3V1***u3V1**)|M
+ vs(**sv**+d3V1***u3V1**)|M-1))
≡ **I**[SM](**f**(**sv**)*(us(**sv**+d1V1***u1V1**)|M
+ vs(**sv**+d1V1***u1V1**)|M-1
+ us(**sv**+d2V1***u2V1**)|M
+ vs(**sv**+d2V1***u2V1**)|M-1
+ us(**sv**+d3V1***u3V1**)|M
+ vs(**sv**+d3V1***u3V1**)|M-1)).

If
sv+diV1***uiV1** is in V3-(M∪SM)
then us(**sv**+diV1***uiV1**|M)+ vs(**sv**+diV1***uiV1**|M)-1 = 1.

If
sv+diV1***uiV1** is in M-SM
then us(**sv**+diV1***uiV1**|M)+ vs(**sv**+diV1***uiV1**|M)-1 = -1.

If
sv+diV1***uiV1** is in SM
then us(**sv**+diV1***uiV1**|M)+ vs(**sv**+diV1***uiV1**|M)-1 = 0.

Einstein: "Now continue for special functions."

Breton: "Let f be continuous, bounded and asymptotic over V3
I[V3]•(vs(v)|M –us(v)|M))*(D[v]*(f(v);dV1))
•[UV1]$^{-1}$•T[UV1]$^{-1}$;dV1)
= I[SM]•(D[v]*(f(v);dV1)•[UV1]$^{-1}$•T[UV1]$^{-1}$;dV1)
= lim S[bn(SM∩Pn)]
(f(sv(Pn(i,j,k))+d1V1*u1V1)–f(sv(Pn(i,j,k)))
+ f(sv(Pn(i,j,k))+d2V1*u2V1)–f(sv(Pn(i,j,k)))
+ f(sv(Pn(i,j,k))+d3V1*u3V1)–f(sv(Pn(i,j,k))))
= 0
since f is continuous.
Also,
I[V3]•((D[v]*(f(v)*us(v)|M;dV1)+D[v]*(f(v)*vs(v)|M;dV1))
•[UV1]$^{-1}$•T[UV1]$^{-1}$;dV1)
= lim S[bn(V3∩Pn)]((D[v(Pn(i,j,k))]*(f(v)*us(v)|M;dV1)
+ D[v(Pn(i,j,k))]*(f(v)*vs(v)|M;dV1))
•[UV1]$^{-1}$•T[UV1]$^{-1}$•m(Pn(i,j,k)))
= lim(S[bn(V3∩Pn)]((f(v(Pn(i,j,k))+d1V1*u1V1)
*us(v+d1V1*u1V1)|M
–f(v(Pn(i,j,k)))*us(v(Pn(i,j,k)))|M
+ f(v(Pn(i,j,k))+d1V1*u1V1)
*vs(v+d1V1*u1V1)|M
–f(v(Pn(i,j,k)))*vs(v(Pn(i,j,k)))|M)
*u1V1/d1V1
+ (f(v(Pn(i,j,k))+d2V1*u2V1)
*us(v+d2V1*u2V1)|M
–f(v(Pn(i,j,k)))*us(v(Pn(i,j,k)))|M
+ f(v(Pn(i,j,k))+d2V1*u2V1)
*vs(v+d2V1*u2V1)|M
–f(v(Pn(i,j,k)))*vs(v(Pn(i,j,k)))|M)
*u2V1/d2V1
+ (f(v(Pn(i,j,k))+d3V1*u3V1)
*us(v+d3V1*u3V1)|M
–f(v(Pn(i,j,k)))*us(v(Pn(i,j,k)))|M
+ f(v(Pn(i,j,k))+d3V1*u3V1)
*vs(v+d3V1*u3V1)|M
–f(v(Pn(i,j,k)))*vs(v(Pn(i,j,k)))|M)
*u3V1/d3V1)
•[UV1]$^{-1}$•T[UV1]$^{-1}$
•(p1V1*u1V1 + p2V1*u2V1 +p3V1*u3V1)/n^3
= lim S[bn(SM∩Pn)](f(sv(Pn(i,j,k))+d1V1*u1V1)
*(us(sv+d1V1*u1V1)|M
+ vs(sv+d1V1*u1V1)|M)
– f(sv(Pn(i,j,k)))
+ f(sv(Pn(i,j,k))+d2V1*u2V1)
*(us(sv+d2V1*u2V1)|M
+ vs(sv+d2V1*u2V1)|M)
–f(sv(Pn(i,j,k)))

+ f(v(Pn(i,j,k))+d3V1*u3V1)
*us(v+d3V1*u3V1)|M
+ vs(sv+d3V1*u3V1)|M)
-f(sv(Pn(i,j,k)))
≡ I[SM](f(sv(Pn(i,j,k))+d1V1*u1V1)
*(us(sv+d1V1*u1V1)|M
+ vs(sv+d1V1*u1V1)|M)
- f(sv(Pn(i,j,k)))
+ f(sv(Pn(i,j,k))+d2V1*u2V1)
*(us(sv+d2V1*u2V1)|M
+ vs(sv+d2V1*u2V1)|M)
-f(sv(Pn(i,j,k)))
+ f(v(Pn(i,j,k))+d3V1*u3V1)
*us((v+d3V1*u3V1)|M
+ vs(sv+d3V1*u3V1)|M)
-f(sv(Pn(i,j,k)))).

This is the **fundamental theorem of integral calculus over measurable sets in V3**."

Einstein, impressed, but determined to press forward: "Other integral operations are possible."

Breton: "Just as with directional gradients."

I[V3]•(f(v)*(D[v]*(us(v)|M;dV1) + D[v]*(vs(v)|M;dV1))
•[UV1]$^{-1}$•T[UV1]$^{-1}$;dV1)
= I[SM](f(sv)
*(u(sv+d1V1*u1V1|M) + v(sv+d1V1*u1V1)|M-1
+ u(sv+d2V1*u2V1)|M + v(sv+d2V1*u2V1)|M-1
+ u(sv+d3V1*u3V1)|M + v(sv+d3V1*u3V1)|M-1))

I[V3]•(f(v)∧(D[v]*(us(v)|M;dV1) + D[v]*(vs(v)|M;dV1))
•[UV1]$^{-1}$•T[UV1]$^{-1}$;dV1)
= lim S[bn(SM∩Pn)](f(sv(Pn(i,j,k)))
•((us(sv+d1V1*u1V1)|M
+ vs(sv+d1V1*u1V1)|M-1)*u1V1/d1V1
+ (us(sv+d2V1*u2V1)|M
+ vs(sv+d2V1*u2V1)|M-1)*u2V1/d2V1
+ (us(sv+d3V1*u3V1)|M
+ vs(sv+d3V1*u3V1)|M-1)*u3V1/d3V1)
•(p1V1*u2V1∧u3V1
+ p2V1*u3V1∧u1V1
+ p3V1*u1V1∧u2V1))

= lim S[bn(SM∩Pn)](f(sv(Pn(i,j,k)))
•((us(sv+d1V1*u1V1)|M
+ vs(sv+d1V1*u1V1)|M−1)
*u1V1∧(u2V1∧u3V1)
+ (us(sv+d2V1*u2V1)|M
+ vs(sv+d2V1*u2V1)|M−1)
*u2V1∧(u3V1∧u1V1)
+ (us(sv+d3V1*u3V1)|M
+ vs(sv+d3V1*u3V1)|M−1)
*u3V1∧(u1V1∧u2V1))
= I[SM](f(sv)
•((us(sv+d1V1*u1V1)|M
+ vs(sv+d1V1*u1V1)|M−1)
*u1V1∧(u2V1∧u3V1)
+ (us(sv+d2V1*u2V1)|M
+ vs(sv+d2V1*u2V1)|M−1)
*u2V1∧(u3V1∧u1V1)
+ (us(sv+d3V1*u3V1)|M
+ vs(sv+d3V1*u3V1)|M−1)
*u3V1∧(u1V1∧u2V1)))

I[V3]•(f(v)∧(D[v]*(us(v)|M;dV1)
+ D[v]*(vs(v)|M;dV1));dV1)
= I[SM](f(sv)
•C((us(sv+d1V1*u1V1)|M
+ vs(sv+d1V1*u1V1)|M−1)*u1V1/d1V1
+ (us(sv+d2V1*u2V1)|M
+ vs(sv+d2V1*u2V1)|M−1)*u2V1/d2V1
+ (us(sv+d3V1*u3V1)|M
+ vs(sv+d3V1*u3V1)|M−1)*u3V1/d3V1)
•pV1)

I[V3]∧(f(v)∧(D[v]*(us(v)|M;dV1)
+ D[v]*(vs(v)|M;dV1));dV1)
= I[SM](f(sv)
•C((us(sv+d1V1*u1V1)|M
+ vs(sv+d1V1*u1V1)|M−1)*u1V1/d1V1
+ (us(sv+d2V1*u2V1)|M
+ vs(sv+d2V1*u2V1)|M−1)*u2V1/d2V1
+ (us(sv+d3V1*u3V1)|M
+ vs(sv+d3V1*u3V1)|M−1)*u3V1/d3V1)
•C(pV1))

```
I[V3]∧(f(v)*(D[v]*(us(v)|M;dV1)
                        + D[v]*(vs(v)|M;dV1));dV1)
        = I[SM](f(sv)
               *((us(sv+d1V1*u1V1)|M
                        + vs(sv+d1V1*u1V1)|M-1)*u1V1/d1V1
               + (us(sv+d2V1*u2V1)|M
                        + vs(sv+d2V1*u2V1)|M-1)*u2V1/d2V1
               + (us(sv+d3V1*u3V1)|M
                        + vs(sv+d3V1*u3V1)|M-1)*u3V1/d3V1)
                                                  •C(dV1))
I[V3]*(f(v)*(D[v]*(us(v)|M;dV1)
                        + D[v]*(vs(v)|M;dV1));dV1)
        = I[SM](f(sv)
               *((us(sv+d1V1*u1V1)|M
                        + vs(sv+d1V1*u1V1)|M-1)*u1V1/d1V1
               + (us(sv+d2V1*u2V1)|M
                        + vs(sv+d2V1*u2V1)|M-1)*u2V1/d2V1
               + (us(sv+d3V1*u3V1)|M
                        + vs(sv+d3V1*u3V1)|M-1)*u3V1/d3V1)
                                                  *dV1)
I[V3]*(f(v)•(D[v]*(us(v)|M;dV1)
                        + D[v]*(vs(v)|M;dV1));dV1)
        = I[SM](f(sv)
               •((us(sv+d1V1*u1V1)|M
                        + vs(sv+d1V1*u1V1)|M-1)*u1V1/d1V1
               + (us(sv+d2V1*u2V1)|M
                        + vs(sv+d2V1*u2V1)|M-1)*u2V1/d2V1
               + (us(sv+d3V1*u3V1)|M
                        + vs(sv+d3V1*u3V1)|M-1)*u3V1/d3V1)
                                                  *dV1)
I[V3]*(f(v)∧(D[v]*(us(v)|M;dV1)
                        + D[v]*(vs(v)|M;dV1));dV1)
        = I[SM](f(sv)
               ∧((us(sv+d1V1*u1V1)|M
                        + vs(sv+d1V1*u1V1)|M-1)*u1V1/d1V1
               + (us(sv+d2V1*u2V1)|M
                        + vs(sv+d2V1*u2V1)|M-1)*u2V1/d2V1
               + (us(sv+d3V1*u3V1)|M
                        + vs(sv+d3V1*u3V1)|M-1)*u3V1/d3V1)
                                                  *dV1)
```

I[V3]*(f(v)*(D[v]*(us(v)|M;dV1)
+ D[v]*(vs(v)|M;dV1));dV1)
= I[SM](f(sv)
*((us(sv+d1V1*u1V1|M)
+ vs(sv+d1V1*u1V1|M)-1)*u1V1/d1V1
+ (us(sv+d2V1*u2V1|M)
+ vs(sv+d2V1*u2V1|M)-1)*u2V1/d2V1
+ (us(sv+d3V1*u3V1|M)
+ vs(sv+d3V1*u3V1|M)-1)*u3V1/d3V1)
*dV1).

For **f** or f continuous, bounded and asymptotic over V3

I[V3]•((vs(v)|M-us(v)|M)*(D[v]*(f(v);dV1))
•[UV1]$^{-1}$•T[UV1]$^{-1}$;dV1)
= I[SM]•(D[vs]*(f(v);dV1)•[UV1]$^{-1}$•T[UV1]$^{-1}$;dV1)
= lim (f(sv+d1V1*u1V1)-f(sv)
+ f(sv+d2V1*u2V1)-f(sv)
+ f(sv+d3V1*u3V1)-f(sv))
= 0

I[V3]•((vs(v)|M-us(v)|M)*(D[v]*f(v);dV1))
•[UV1]$^{-1}$•T[UV1]$^{-1}$;dV1)
= 0

I[V3]•((vs(v)|M-us(v)|M)*(D[v]∧(f(v);dV1));dV1)
= I[SM](D[sv]∧(f(v);dV1)•dV1)

I[V3]∧((vs(v)|M-us(v)|M)*(D[v]∧(f(v);dV1));dV1)
= I[SM](D[v]∧(f(v);dV1)∧dV1)

I[V3]∧((vs(v)|M-us(v)|M)*(D[v]*(f(v);dV1));dV1)
= I[SM](D[v]*(f(v);dV1)∧dV1)

I[V3]*((vs(v)|M-us(v)|M)*(D[v]•(f(v);dV1));dV1)
= I[SM](D[v]•(f(v);dV1)*dV1)

I[V3]*((vs(v)|M-us(v)|M)*(D[v]∧(f(v);dV1));dV1)
= I[SM](D[v]∧(f(v);dV1)*dV1)

I[V3]*((vs(v)|M-us(v)|M)*(D[v]*(f(v);dV1));dV1)
= I[SM](D[v]*(f(v);dV1)*dV1).

Newton: "We're sprouting equations. Surfaces are appearing as central issue. Please focus on them"

Breton: "let's first talk about differentiable surfaces. A differentiable surface is defined by two parameters **sv** = **v**(x1,x2) where

D(**sv**|x2;x1) and D(**sv**|x1;x2)

are basic derivatives at **vs1**.

If the surface of a set M is differentiable at **sv1**, there exists at **sv1** three well-defined orthogonal vectors: two tangential unit vectors **ut1**(**sv1**),and **ut2**(**sv1**), and an orthogonal vector **un**(**sv1**). For **un** constructed to be the "outward" orthogonal into **V3**-(M∪SM) this local orientation may be used to define both directional and sectional gradients."

Derivatives and Integrals over Surfaces

Einstein: "Continue."

Breton: "Differentiable surfaces at **sv1** may be classified as following:

sv1+d1***ut1** in (M–SM) or in **V3** - (MUSM)
sv1+d2***ut2** in (M–SM) or in **V3** - (MUSM)
sv1–d1***ut1** in (M–SM) or in **V3** - (MUSM)
sv1–d2***ut2** in (M–SM) or in **V3** - (MUSM)
sv1+d3***un** in (M–SM) or in **V3** - (MUSM)

Directional and sectional gradients may then be defined according a local system of quadrant gradients. There are 32 classifications of such differentiable surfaces. For each classification eight local quadrant gradients **D[sv1]***(us(**v**|M);**dVn**) may be defined."

Einstein: "That's a big mouthful. Give us an example."

Breton: "OK."

Example. Let **vs1** be an element of the surface. Let **vs1**+d1***ut1**, **vs1**–d1***ut1**, and **vs1**–d2***ut2** all lie in SM for d1 and d2 positive and sufficiently small. Let **vs1**+d2***ut2** and **vs1**+d3***un** lie in **V3**–(MUSM) for d2 and d3 positive and sufficiently small. For this classification,

D[**sv1**]*(us(**v**|M);**dV1n**) ≡ us(d1***ut1**)***ut1**/d1
+ us(d2***ut2**)***ut2**/d2
+ us(d3***un**)***un**/d3
= **qd**(d2***ut2**) + **qd**(d3***un**)

D[**sv1**]*(us(**v**|M);**dV2n**) ≡ (us(–d1***ut1**))***ut1**/d1
+ (us(+d2***ut2**))***ut2**/d2
+ (us(+d3***un**))***un**/d3
= **qd**(d2***ut2**) + **qd**(d3***un**)

D[**sv1**]*(us(**v**|M);**dV3n**) ≡ (us(+d1***ut1**))***ut1**/d1
+ (us(–d2***ut2**))***ut2**/d2
+ (us(+d3***un**))***un**/d3
= **qd**(d3***un**)

D[**sv1**]*(us(**v**|M);**dV4n**) ≡ (us(–d1***ut1**))***ut1**/d1
+ (us(–d2***ut2**))***ut2**/d2
+ (us(+d3***un**))***un**/d3
= **qd**(d3***un**)

D[**sv1**]*(us(**sv1**|M);**dV5n**) ≡ (us(+d1***ut1**))***ut1**/d1
+ (us(+d2***ut2**))***ut2**/d2
+ (us(–d3***un**))***un**/d3
= **qd**(d2***ut2**)

D[**sv1**]*(us(**v**|M);**dV6n**) ≡ (us(−d1***ut1**))***ut1**/d1
+ (us(+d2***ut2**))***ut2**/d2
+ (us(−d3***un**))***un**/d3
= **qd**(d2***ut2**)

D[**sv1**]*(us(**v**|M);**dV7n**) ≡ (us(+d1***ut1**))***ut1**/d1
+ (us(−d2***ut2**))***ut2**/d2
+ (us(−d3***un**))***un**/d3
= **0**

D[**sv1**]*(us(**v**|M);**dV8n**) ≡ (us(−d1***ut1**))***ut1**/d1
+ (us(−d2***ut2**))***ut2**/d2
+ (us(−d3***un**))***un**/d3
= **0**.

Einstein: "Let's consider only orthogonals to the surface."

Breton: "Here is a definition."

Definition (directional derivative and gradients orthogonal to a surface)

Given
- M, a measurable set with a differentiable surface SM;
- **sv**, an element of SM;
- **un(sv)**, the outward unit orthogonal vector of M at **sv**;

for
- **f**, a function with a basic derivative at **sv** in direction **un(sv)**;

lim(**f(sv**+d***un(sv**))−f(**sv**))/d as d → 0

is called the **basic directional derivative orthogonal to** M **at sv**

lim(**f(sv**+d***un(sv**))−**f(sv**))***qd**(d***un(sv**)) as d → 0

is called the **basic directional gradient orthogonal to M at sv**.

end of definition

Orthogonal directional derivatives are written
D[**sv,sv+dun(sv)**](**f(v)**|M;dv).
Orthogonal directional gradients are written
D[**sv,sv+dun(sv)**]*(**f(r)**|M;**dun(sv)**).

Newton: "These are just specialized basic derivatives."

Breton: "Just what we should expect."

Einstein: "How about step functions?"

Breton: "The orthogonal unit point step functions are written

$$\mathrm{us}(\mathbf{v}-\mathbf{sv}\,|\,\mathbf{un}(\mathbf{sv}))$$

and

$$\mathrm{vs}(\mathbf{v}-\mathbf{sv}\,|\,\mathbf{un}(\mathbf{sv}))$$

with accompanying gradients written as

$$\mathbf{D}[\mathbf{sv},\mathbf{sv}+\mathbf{dun}(\mathbf{sv})]*(\mathrm{us}(\mathbf{v}-\mathbf{sv})\,|\,M;\mathbf{dun}(\mathbf{sv}))$$

and

$$\mathbf{D}[\mathbf{sv},\mathbf{sv}+\mathbf{dun}(\mathbf{sv})]*(\mathrm{vs}(\mathbf{v}-\mathbf{sv})\,|\,M;\mathbf{dun}(\mathbf{sv})).$$

Einstein: "And how are functions implicated?"

Breton: "Now for **f** continuous, bounded, and asymptotic

$$\mathrm{I}[\mathbf{V3}](\mathrm{us}(\mathbf{v}-\mathbf{sv}\,|\,\mathbf{un}(\mathbf{sv}))*(\mathbf{D}[\mathbf{v},\mathbf{v}+\mathbf{dun}(\mathbf{sv})]\bullet(\mathbf{f}(\mathbf{v})\,|\,\mathbf{un}(\mathbf{sv});\mathbf{dv})) \\ ;\mathrm{abs}(\mathbf{dun}(\mathbf{sv}))) \\ = \lim\,(\mathbf{f}(\mathbf{sv}+d*\mathbf{un}(\mathbf{sv}))\bullet\mathbf{un}(\mathbf{sv})/d)*\mathrm{abs}(d*\mathbf{un}(\mathbf{sv})) \\ = \mathbf{f}(\mathbf{sv})\bullet\mathbf{un}(\mathbf{sv})$$

and

$$\mathrm{I}[\mathbf{V3}]((1-\mathrm{us}(\mathbf{v}-\mathbf{sv}\,|\,\mathbf{un}(\mathbf{sv})))*\mathbf{D}[\mathbf{v},\mathbf{v}+\mathbf{dun}(\mathbf{sv})]\bullet(\mathbf{f}(\mathbf{v})\,|\,\mathbf{un}(\mathbf{sv});\mathbf{dv}) \\ ;\ \mathrm{abs}(\mathbf{dun}(\mathbf{sv}))) \\ = -\mathbf{f}(\mathbf{sv})\bullet\mathbf{un}(\mathbf{sv}).$$

A similar integration may be performed for each element of SM."

Einstein: "How about sectional gradients?"

Breton: "A section need not be oriented orthogonally to a surface. We have discussed sectional gradients generally. What more do you want?"

Einstein: "I can imagine a section oriented orthogonally. It would contain **un**."

Breton: "There may be several such sections. Let me define one such together with the gradients which can be associated with it.

Definition (sectional gradients orthogonal to a surface)

Given

M, a measurable set with a differentiable surface SM;

sv, an element of SM;

ut1(sv), **ut2(sv)**, **un(sv)**,
orthogonal and tangent vectors of M at **sv**;

for

$$\mathbf{us1}(\mathbf{sv}) = (\mathbf{un}(\mathbf{sv}) + \mathbf{ut1}(\mathbf{sv}) + \mathbf{ut2}(\mathbf{sv}))/\sqrt{3}$$

$$\mathbf{us2}(\mathbf{sv}) = (2*\mathbf{un}(\mathbf{sv}) - (\sqrt{3}+1)*\mathbf{ut1}(\mathbf{sv}) \\ + (\sqrt{3}-1)*\mathbf{ut2}(\mathbf{sv}))/(2*\sqrt{3})$$

$$\mathbf{us3}(\mathbf{sv}) = (2*\mathbf{un}(\mathbf{sv}) + (\sqrt{3}-1)*\mathbf{ut1}(\mathbf{sv}) \\ - (\sqrt{3}+1)*\mathbf{ut2}(\mathbf{sv}))/(2*\sqrt{3})$$

f a function with a basic sectional gradient at **sv**

---

then

lim (**f**(**sv**+d1***us1**(**sv**))-**f**(**sv**))***us1**(**sv**)/d1
+ (**f**(**sv**+d2***us2**(**sv**))-**f**(**sv**))***us2**(**sv**)/d2
+ (**f**(**sv**+d3***us3**(**sv**))-**f**(**sv**))***us3**(**sv**)/d3
sv in SM
is called the **basic orthogonal gradient to** M **at sv**.
end of definition

For SECT(**sv**,**us1**(**sv**),**us2**(**sv**),**us3**(**sv**)) orthogonal gradients are written as

D[**sv**]*(**f**(**v**) | M;**dVn**).

Einstein: "That explains **dVn**, but how do you arrive at this definition?"

Breton: "The sectional axes (**us1**, **us2**, **us3**) is the coordinate system (**un**,**ut1**,**ut2**) rotated such that (**us1**+**us2**+**us3**)/sqrt(3) = **un**. You can use your acquired algebraic skills to see that this is so."

Einstein: "Please illustrate."

Breton: "As illustration, suppose the local surface planar at **sv1** with **sv1**+d1***ut1** and **sv1**+d2***ut2** lie in a planar surface. Then

D[sv1]*(us(**v**|M);**dVn**) = lim **un**/d3
= **D[sv1,sv1+dVn]***(us(**v-sev**)|M;**dVn**).

Contrariwise,

D[sv1]*(vs(**v**|M);**dVn**) = **0**.

In the "inward" direction -**un** into M-SM

D[sv1]*(us(**v**|M);**dVn**$^-$) = **0**

whereas,
D[sv1]*(vs(**v**|M);**dVn**$^-$)
= lim **un**/d3
=-**D[sv1,sv1-dVn**$^+$**]***(vs(**v-sev**)|M;**dVn**)
D[**sv**]*(us(**v**|M)|M;**dVn**)
= lim (**us1**/d1 + **us2**/d2 + **us3**/d3)
D[**sv**]*(vs(**v**|M)|M;**dVn**) = **0**.

Einstein: "Can we bring local step functions into the picture?"

Breton: "An analogous development can be made for local step functions.
Let **u1+u2+u3** be a partition vector and **f** continuously locally differentiable with a basic gradient and bounded over a measurable set M with differentiable surface SM at each point **sv** of which are defined a set of orthogonal directions **ut1**(**sv**), **ut2**(**sv**), and **un**(**sv**).

Then because of interior cancellation,

I[M](D[**v**])•(**f**(**v**);**dv**)*abs(**dv**)

has values only at the surface. Each cell in Pn at the surface may lie partly in M and partly outside of M. The interior part of the cell undergoes interior cancellation, while the exterior remains to be summed. Here I'll sketch our what I mean."

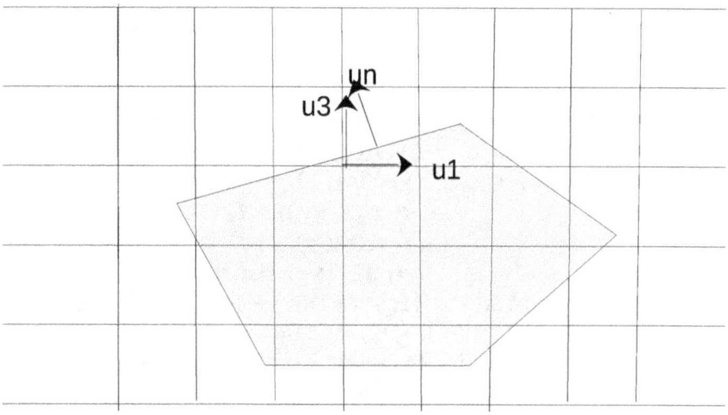

Einstein: "So in any partition, only the interior part of the surface cell is subject to interior cancellation. This circumstance will lead to error."

Breton: "The fraction of the cell corresponding to **f**(**sv**(Pn(i,j,k))+d***ui**) is

$((\mathbf{ut1} \wedge \mathbf{ut2}) \bullet \mathbf{ui} / n^3)/n^3 = (\mathbf{ut1} \wedge \mathbf{ut2}) \bullet \mathbf{ui}$

and thus invariant with n."

Einstein: "Where does this come from?"

Breton: "Consider a cell in the nth partition of M with a partition vector **u1+u2+u3**. What is its volume?"

Newton: "Each cell is a cube with one side equal to 1/n. So one cell in the Pn partition has a volume of $1/n^3$."

Breton: "Now tell me what is the volume of the corresponding orthogonal surface section specified by (**ut1**, **ut2**, **ui**)?"

Newton: "That is just the triple product.$(\mathbf{ut1} \wedge \mathbf{ut2}) \bullet \mathbf{ui}$."

Breton: "In the nth partition?"

Newton: "That would be $((\mathbf{ut1} \wedge \mathbf{ut2}) \bullet \mathbf{ui} / n^3)$."

Breton: "So the ratio of the two would be

$$((\mathbf{ut1} \wedge \mathbf{ut2}) \bullet \mathbf{ui} / n^3)/n^3$$

Newton: "So the n's cancel leaving only $(\mathbf{ut1} \wedge \mathbf{ut2}) \bullet \mathbf{ui}$."

Breton: "For any partition even in the limit."

Einstein: "And how does this affect integration."

Breton: "Applying these weights to the cells at the surface

I[M]•(D[**v**]*(**f**(**v**);**dv**);**dv**)
= lim S[bn(SM∩Pn)](**f**(**sv**(Pn(i,j,k))+d***u1**)
*(**ut1**(**sv**)∧**ut2**(**sv**))•**u1**
+ **f**(**sv**(Pn(i,j,k))+d***u2**)
*(**ut1**(**sv**)∧**ut2**(**sv**))•**u2**
+ **f**(**sv**(Pn(i,j,k))+d***u3**)
*(**ut1**(**sv**)∧**ut2**(**sv**))•**u3**)
= lim S[bn(SM∩Pn)](**f**(**sv**(Pn(i,j,k))+d***u1**)***u1**•**un**(**sv**)
+ **f**(**sv**(Pn(i,j,k))+d***u2**)***u2**•**un**(**sv**)
+ **f**(**sv**(Pn(i,j,k))+d***u3**)***u3**•**un**(**sv**))
= **I**[SM](**f**(**sv**)*(**u1**+**u2**+**u3**)•**un**(**sv**))
I[M]•(D[**v**]∧(**f**(**v**);**dv**);**dv**) = 0
I[M]∧(D[**v**]∧(**f**(**v**);**dv**);**dv**)
= **I**[SM]([**u1*****u1** – **I**]•**f**(**sv**)***u1**
+[**u2*****u2** – **I**]•**f**(**sv**)***u2**
+[**u3*****u3** – **I**]•**f**(**sv**)***u3**)•**un**(**sv**)
I[M]*(D[**v**]•(**f**(**v**);**dv**);**dv**)
= **I**[SM]([**u1*****f**(**sv**)•[**u1*****u1**]
+ [**u2*****f**(**sv**)]•[**u2*****u2**]
+ [**u3*****f**(**sv**)]•[**u3*****u3**]))
I[M]*(D[**v**]]∧(**f**(**v**);**dv**);**dv**)
= **I**[SM](**C**(**f**(**sv**))•([**u1*****u1**]*(**u1**•**un**)
+ [**u2*****u2**]*(**u2**•**un**)
+ [**u3*****u3**]*(**u3**•**un**))).

Einstein: " We've defined a number of step functions. It would be helpful to have a table of the different kinds."

Newton: "I concur. In fact, I have been tabulating just such a table and here it is.

Step Function	Description	Stepping point
u(q–q1)	over Q	in Q
us(q–q1)	over **v**(q)	in **V3**
us(q - qi)|**uv**	over a direction.	in a directional line
us(**v**(q) - **v**(qi))|CQ	over a curve	in a curve. **v**(q)
us(**v** - **vi**)|CV	over a curve	in a curve
us(**v**–**sev**|M)	over **V3**	in a surface element
us(**v**)|M	over **V3**	in the whole surface
us(**v**–**sv**|**un**(**sv**))	over **V3**	surface orthogonal

Newton: "The table give only the 'us' step function. The 'vs' step function can be easily be reconstructed similarly."

The Divergence Theorem and Green's Theorem

Breton: "A variation of these results forms the foundation of the classical divergence theorem."

Einstein: "First explain the divergence theorem."

Breton: "One of Gauss' masterpieces, the divergence theorem is expressed by a three dimensional integration using a volumetric non–vector measure v ≈ **u1**•(**u2**∧**u3**) and a second non–vector measure for an integration related to surface. The theorem states

I[M](D•**f**;dv) = I[SM](**f**•**un**;ds)

where **un** is the outward orthogonal on the surface designated by SM."

Einstein: "An important theorem indeed. I have noted its use in many books on Physics."

Breton: "We have climbed up sufficiently on our mountainous quest to be able to appraise the divergence theorem."

Einstein: "What do you mean?"

Breton: "We can enumerate when the divergence theorem fails. If,

- **f**(**v**) is discontinuous in M
- **f**(**v**) is unbounded in M
- D•**f**; is not basic or is discontinuous
- M is non–measurable
- SM is non–differentiable."

Einstein: "And how do your equations correspond to the divergence theorem?"

Breton: "Like this

```
I[M](D[v)]•(f(v);dv)*abs(dv)
        = lim S[bn(SM∩Pn)](f(sv(Pn(i,j,k))+d*u1)•u1
                        *(ut1(sv)∧ut2(sv))•u1
                  + f(sv(Pn(i,j,k))+d*u2)•u2
                        *(ut1(sv)∧ut2(sv))•u2
                  + f(sv(Pn(i,j,k))+d*u3)•u3)
                        *(ut1(sv)∧ut2(sv))•u3)
        = lim S[bn(SM∩Pn)]
                  (f(sv(Pn(i,j,k))+d*u1)•u1*u1•un(sv)
                  + f(sv(Pn(i,j,k))+d*u2)•u2*u2•un(sv)
                  + f(sv(Pn(i,j,k))+d*u3)•u3*u3•un(sv))
        =I[SM](f(sv)•un(sv)).
```

Physicists often use another theorem, Green's theorem,
I[M]((f*D•D*g–g*D•D*f);dv) = I[SM](f*D*g–g*D*f)•un;ds)
which suffers the same restrictions as the divergence theorem."

Einstein: "Show us."

Breton: "Let f(v) and g(v) be locally differentiable scalar functions with basic gradients and bounded over M. Then

```
I[M](D[v]•(f(v)*D[v]*(g(v);dv)
                  –g(v)*D[v]*(f(v);dv);dv);abs(dv))
        = I[M]((f(v)*D[v]•(D[v]*(g(v);dv);dv)
                  + D[v]*(f(v);dv)•D[v]*(g(v);dv)
                        –g(v)*D[v]•(D[v]*(f(v);dv);dv)
                  –D[v]*(f(v);dv)•D[v]*(g(v);dv));abs(dv))
        = I[M]((f(v)*D[v]•(D[v]*(g(v);dv);dv)
                        –g(v)*D[v]•(D[v]*(f(v);dv);dv));abs(dv))
        = I[SM](f(sv)*D[sv]*(g(v);dv)
                        –g(v)*D[sv]*(f(v);dv))•un(sv).
```

Newton: "Variations may likewise be formulated for curls, gradients, invergences, incurls and ingradients?"

Breton: "Yes. For vector functions **f(v)** and **g(v)** under the same conditions

```
I[M]((D[v]∧(g(v);dv)•T[(D[v]*(f(v);dv);dv));abs(dv))
        = I[SM](f(sv)*D[sv]∧(g(v);dv))•un(sv).
```

Einstein: "So what we have developed comprehends both the divergence and the Green's theorem."

Breton: "But again the surface may not be differentiable at **sv**. Such cases may still be analyzed by sectional gradients. Our equations, besides allowing analysis of cases for which the divergence theorem fails, may also be used to formulate analogs to the divergence theorem for curls and gradients. Analysis by sectional gradients embraces quadrant gradients. Sectional gradients of us((**v**|M)) may be used generally not only for differentiable surfaces, but also those with angles, corners, etc. Where the surface becomes an inner cusp, for instance, the sectional gradient may even become a directional gradient."

Newton: "Explain this further."

Breton: "A differentiable surface is defined by unique orthogonals at every point in the surface. Non-differentiable surfaces, however, may have more than one local orthogonal. For instance where the surface at **v1** is the conjunction of many planes, a section can be assigned to each plane with **u1Vi** and **u2Vi** co planar and **u3Vi** orthogonal. The surface then at **vs1** would have multiple orthogonals. In the sequel we assume surfaces with no more than a denumerable number of such irregularities; results are stated in terms of sectional gradients."

Einstein: "Give us an example."

Breton: "Here is one which contrasts the two approaches.

Example. Let $\mathbf{f=v-sv0}$ and SM be a differentiable surface containing $\mathbf{sv0}$ which supports an orthogonal orientation.

$$\mathbf{D}[\mathbf{v}]*(\mathbf{f};\mathbf{dv}) = \mathbf{I}$$
$$\mathbf{D}[\mathbf{v}]\wedge(\mathbf{f};\mathbf{dv}) = \mathbf{0}$$
$$\mathrm{D}[\mathbf{v}]\bullet(\mathbf{f};\mathbf{dv}) = 3.$$

Sectionally for $\mathbf{dV1}$

$$\mathrm{D}[\mathbf{v}]*(\mathbf{f};\mathbf{dV1}) = \mathbf{u1V1*u1V1 + u2V1*u2V1 + u3V1*u3V1}$$
$$\mathrm{D}[\mathbf{v}]\wedge(\mathbf{f};\mathbf{dV1}) = \mathbf{0}$$
$$\mathrm{D}[\mathbf{v}]\bullet(\mathbf{f};\mathbf{dV1}) = 3.$$

Directionally for $\mathbf{uv}$

$$\mathrm{D}[\mathbf{v,v+dv}]*(\mathbf{f}\,|\,\mathbf{uv};\mathbf{dv}) = \mathbf{uv*uv}$$
$$\mathrm{D}[\mathbf{v,v+dv}]\wedge(\mathbf{f}\,|\,\mathbf{uv};\mathbf{dv}) = \mathbf{0}$$
$$\mathrm{D}[\mathbf{v,v+dv}]\bullet(\mathbf{f}\,|\,\mathbf{uv};\mathbf{dv}) = 1$$
$$\mathbf{I}[\mathbf{v1,v2}]*(\mathbf{f}\,|\,\mathbf{uv};\mathbf{dv}) = (\mathbf{v2*v2-v1*v1 +v1*v2-v2*v1})/\mathbf{2} \\ -\mathbf{sv0}*(\mathbf{v2-v1})$$
$$\mathbf{I}[\mathbf{v1,v2}]\wedge(\mathbf{f}\,|\,\mathbf{uv};\mathbf{dv}) = (\mathbf{v1-sv0})\wedge(\mathbf{v2-v1})$$
$$\mathbf{I}[\mathbf{v1,v2}]\bullet(\mathbf{f}\,|\,\mathbf{uv};\mathbf{dv}) = (\mathbf{v2\bullet v2-v1\bullet v1})/\mathbf{2} - \mathbf{sv0}\bullet(\mathbf{v2-v1}).$$

For a given measurable set M

$$\mathbf{I}[\mathrm{M}]\bullet(\mathbf{D}[\mathbf{v}]*(\mathbf{f};\mathbf{dv});\mathbf{dv}) = \mathbf{m}(\mathrm{M}) = \mathrm{vol}(\mathrm{M})*(\mathbf{u1+u2+u3})$$
$$\mathbf{I}[\mathrm{M}]*(\mathbf{D}[\mathbf{v}]\bullet(\mathbf{f};\mathbf{dv});\mathbf{dv}) = 3*\mathbf{m}(\mathrm{M})$$
$$\mathbf{I}[\mathbf{V3}]\bullet(\mathbf{f}(\mathbf{v})*\mathbf{D}[\mathbf{v,v+dV1}]*(\mathrm{u}(\mathbf{v-sev})\,|\,\mathrm{M};\mathbf{dV1});\mathbf{dV1}) \\ = \mathbf{sev} - \mathbf{sv0}$$

I[V3] • (f(v)*D[v,v+dV1]*(v(v−sev) | M;dV1);dV1)
= −sev +sv0
I[V3]*(u(v−sev)*D[v,v+dV1] • (f(v);dV1) | M;dV1)
I[V3]*(u(v−sev) | M;dV1), unbounded.
I[V3] • (f(v)*(D[v]*(us(v) | M;dV1)
+ D[v]*(vs(v) | M;dV1)) • [UV1]$^{-1}$ • T[UV1]$^{-1}$;dV1)
= I[SM](v −sv0) *(us(sv+d1R1*u1R1) | M
+ vs(sv+d1R1*u1V1) | M−1
+ us(sv+d2R1*u2V1) | M
+ vs(sv+d2R1*u2V1) | M−1
+ us(sv+d3R1*u3V1) | M
+ vs(sv+d3R1*u3V1) | M−1).
I[V3]*((vs(v) | M −us(v) | M)*D[v] • (f(v);dv);dv)
= 3*I[SM]*(1;dv)
= 0.
I[M] • (D[v]*(f(v);dv);dv)
= I[M] • (I;dv)
= I[SM](sv−sv0)*(u1+u2+u3) • un(sv).

Newton: "Anything else?"

Breton: "It seems we could go on some more. Let's consider the effect on dilated sets."

Einstein: "Please do."

Breton: "Now let **a** = a1***u1** + a2***u2** + a3***u3** be a constant vector with **G(a)** as its diagonal matrix. Then **v**•**G(a)** = **a**•**G(v)** is a **dilation** of **V3**.

Consider now the dilation of the measurable set M symbolized as M•**G(a)** and the step function us(**v** | M•**G(a)**). Then us(**v** | M•**G(a)**) has the same properties in the dilation as us(**v** | M) in the undilated set."

Einstein: "And what are the gradients of this dilated step function and how are they related to the undilated gradients?"

Breton: "Let's calculate.
D[v•G(a)]*(us(v•G(a) | M•G(a));dV1•G(a))
= lim (us((v+d1V1*u1V1)•G(a)) | M•G(a)
−us(v•G(a) | M•G(a))*u1V1•G(a)
/ abs(d1V1*u1V1•G(a)))
+ (us((v+d2V1*u2V1)•G(a)) | M•G(a)
−us(v•G(a) | M•G(a))*u2V1•G(a)
/ abs(d2V1*u2V1•G(a)))
+ (us((v+d3V1*u3V1)•G(a)) | M•G(a)
−us(v•G(a) | M•G(a))*u3V1•G(a)
/ abs(d3V1*u3V1•G(a)))

$$= \lim ((us(\mathbf{v}+d1V1\ast\mathbf{u1V1})|M-us(\mathbf{v})|M)\ast\mathbf{u1V1}$$
$$/ abs(d1V1\ast\mathbf{u1V1}\bullet\mathbf{G(a)})$$
$$+ (us(\mathbf{v}+d2V1\ast\mathbf{u2V1})|M-us(\mathbf{v})|M)\ast\mathbf{u2V1}$$
$$/ abs(d2V1\ast\mathbf{u2V1}\bullet\mathbf{G(a)})$$
$$+ (us(\mathbf{v}+d3V1\ast\mathbf{u3V1})|M-us(\mathbf{v})|M)\ast\mathbf{u3V1}$$
$$/ abs(d3V1\ast\mathbf{u3V1}\bullet\mathbf{G(a)}))\bullet\mathbf{G(a)}$$
$$= \lim ((us(\mathbf{v}+d1V1\ast\mathbf{u1V1})|M-us(\mathbf{v})|M)\ast\mathbf{u1V1}/d1V1$$
$$+ (us(\mathbf{v}+d2V1\ast\mathbf{u2V1})|M-us(\mathbf{v})|M)\ast\mathbf{u2V1}/d2V1$$
$$+ (us(\mathbf{v}+d3V1\ast\mathbf{u3V1})|M-us(\mathbf{v}|)M)\ast\mathbf{u3V1}/d3V1)$$
$$\bullet[\mathbf{UV1}]^{-1}$$
$$\bullet(\mathbf{u1}\ast\mathbf{u1}/abs(\mathbf{u1V1}\bullet\mathbf{G(a)})$$
$$+ \mathbf{u2}\ast\mathbf{u2}/abs(\mathbf{u2V1}\bullet\mathbf{G(a)})$$
$$+ \mathbf{u3}\ast\mathbf{u3}/abs(\mathbf{u3V1}\bullet\mathbf{G(a)}))$$
$$\bullet[\mathbf{UV1}]\bullet\mathbf{G(a)}.$$

Thus

$$\mathbf{D}[\mathbf{v}\bullet\mathbf{G(a)}]\ast(us(\mathbf{v}\bullet\mathbf{G(a)})|M\bullet\mathbf{G(a)};\mathbf{dV1}\bullet\mathbf{G(a)})$$
$$= \mathbf{D}[\mathbf{v}]\ast(us(\mathbf{v}|M);\mathbf{dV1})\bullet[\mathbf{UV1}]^{-1}$$
$$\bullet(\mathbf{u1}\ast\mathbf{u1}/abs(\mathbf{u1V1}\bullet\mathbf{G(a)})$$
$$+ \mathbf{u2}\ast\mathbf{u2}/abs(\mathbf{u2V1}\bullet\mathbf{G(a)})$$
$$+ \mathbf{u3}\ast\mathbf{u3}/abs(\mathbf{u3V1}\bullet\mathbf{G(a)}))$$
$$\bullet[\mathbf{UV1}]\bullet\mathbf{G(a)}.$$

Likewise,

$$\mathbf{D}[\mathbf{v}\bullet\mathbf{G(a)}]\ast(vs(\mathbf{v}\bullet\mathbf{G(a)})|M\bullet\mathbf{G(a)};\mathbf{dV1}\bullet\mathbf{G(a)})$$
$$= \mathbf{D}[\mathbf{v}]\ast(vs(\mathbf{v}|M);\mathbf{dV1})\bullet[\mathbf{UV1}]^{-1}$$
$$\bullet(\mathbf{u1}\ast\mathbf{u1}/abs(\mathbf{u1V1}\bullet\mathbf{G(a)})$$
$$+ \mathbf{u2}\ast\mathbf{u2}/abs(\mathbf{u2V1}\bullet\mathbf{G(a)})$$
$$+ \mathbf{u3}\ast\mathbf{u3}/abs(\mathbf{u3V1}\bullet\mathbf{G(a)}))$$
$$\bullet[\mathbf{UV1}]\bullet\mathbf{G(a)}.$$

Newton: "For $\mathbf{a} = a$

$$\mathbf{D}[a\ast\mathbf{v}]\ast(us(a\ast\mathbf{v})|\mathbf{a}\ast M;a\ast\mathbf{dV1})$$
$$= (a/abs(a))\ast\mathbf{D}[\mathbf{v}]\ast(us(\mathbf{v}|)M;\mathbf{dV1})$$
$$\mathbf{D}[a\ast\mathbf{v}]\ast(vs(a\ast\mathbf{v})|a\ast M;a\ast\mathbf{dV1})$$
$$= (a/abs(a))\ast\mathbf{D}[\mathbf{v}]\ast(vs(\mathbf{v}|M);\mathbf{dV1}).$$

Einstein: "So we have delved deeply into the foundations of a calculus of V3."

Breton: "Significantly we have broadened the scope of functions which can be considered. Local functions with finite step discontinuities at points or surfaces may be decomposed as:

f(**v**) = **f0**(**v**) + S[n](**fn**) + S[m](**fm**)

where

f0 has no discontinuities;
fn = **a1n***us(**v**-**vn**) in section V1n
= **a2n***us(**v**-**vn**) in section V2n
= ...
fm = **b1m***us(**v**-**svm**|VSm) in section VS1m
= **b2m***us(**v**-**svm**|VSm) in section VS1m
= ...

Such complicated functions may now be differentiated and integrated."

Newton: "How about some specific coordinates, like spherical or cylindrical coordinate systems?"

Breton: "The above results may be further elaborated in terms of specific coordinate systems, not necessarily orthogonal, in the set of vectors. Well chosen coordinates can often make the solution to a particular problem more apparent; but poorly chosen coordinates can often obscure solutions. In this text, results are expressed generically in terms of the arbitrary orientation of the origin and impose no specific system of coordinates."

Einstein: "We summarized differentiation. Why not a table for integration?"

Breton: "Why not?"

Summary of local integration.

Given bounded functions with measure **m** over a measurable set M of either a curve C or V3, the following statements hold for local integrals whether directional or sectional:

Item:	invergence I[M]•	**incurl** I[M]∧	**ingradient** I[M]*
c constant			c***m**(M)
c constant	**c•m**(M)	**c∧m**(M)	**c*m**(M)
c*f			C*I*f
c***f**	c*I•**f**	c*I∧**f**	c*I***f**
c*f	**c**•I*f	**c**∧I*f	**c***I*f
c•f			c•I*f
c∧f	c•I∧f	**c•(T**[I***f**]-I***f**)	**C(c)**•I***f**
c*f	**c***I•**f**		
f+g			I*f + I*g
f+g	I•f + I•g	I∧f +I∧g	I*f + I*g
D*(f*g)	I•(f***D***g) +I•(g***D***f)	I∧(f***D***g) +I∧(g***D***f)	I*(f***D***g) +I*(g***D***f)
(**D***f)/f^2	-I•(**D***(1/f))	-I∧(**D***(1/f))	-I*(**D***(1/f))
D•(f***g**)			I*(f*D•**g**) +I*(**g**•**D***f)
D∧(f***g**)	I•(f***D**∧**g**) +I•(**g**∧**D***f)	I∧(f***D**∧**g**) +I∧(**g**∧**D***f)	I*(f***D**∧**g**) +I*(**g**∧**D***f)
D*(f***g**)	I•(f***D*****g**) +I•(**g*****D***f)	I∧(f***D*****g**) +I∧(**g*****D***f)	I*(f***D*****g**) +I*(**g*****D***f)
D*(**f•g**)	I•(**f•D*g**) +I•(**g•D*f**)	I∧(**f•D*g**) +I∧(**g•D*f**)	I*(**f•D*g**) +I*(**g•D*f**)
D•(**f**∧**g**)			I*(**g•D∧f**) -I*(**f•D∧g**)

Item:	invergence I[M]•	incurl I[M]∧	ingradient I[M]*
D∧(f∧g)	I•(f•T[D*g] -I•(g•T[D*f]) -I•(f*D•g) +I•(g*D•f)	I∧(f•T[D*g] -I∧(g•T[D*f]) -I∧(f*D•g) +I∧(g*D•f)	I*(f•T[D*g]))) -I*(g•T[D*f]) -I*(f*D•g) +I*(g*D•f)
D*(f∧g)	I•(C(f)•D*g) -I•(C(g)•D*f)	I∧(C(f)•D*g) -I∧(C(g)•D*f)	
D•(f*g)	I•(f*D•g) + I•(g•T[D*f])	I∧(f*D•g) + I∧(g•T[D*f])	I*(f*D•g) + I*(g•T[D*f])

Integrals of Sums and Products

As with the table of local derivatives, this table can be used to elaborate other combinations.

Einstein: "Why not go further and give us a table for the various symbols we have been using?"

Breton: "Why not?"

The following two tables reference the symbolism used to distinguish local derivatives and integrals. The symbols for gradients and ingradients are also extended to divergences, curls, invergences and incurls.

Type	Derivative	Gradient
Directional	D[**v1**,**v1+dv**](**f**(**v**)|**uv**;dq)	**D[v1,v1+dv]*(f(v)|uv;dv)**
Along **v(q)**		
with respect to q	D[q,q+dq](**f**(q)|CQ;dq)	
with respect to **v**	D[**v1**,**v1+dv**](**f**(**v**)|CV;dv)	**D[v1,v1+dv]*(f(v)**|CV;**dv)**
positive quadrant		**D[v1]•(f(v);dv)**
Sectional		**D[v1]*(f;dV1)**

Symbols for local Gradients

Type	Integral	Ingradient
Directional	I[**v1**,**v2**](**f**(**v**)|**uv**;dv)	I[**v1**,**v2**]∗(**f**(**v**)|**uv**;**dv**)
Along **v(q)**		
with respect to q	I[q1,q2](**f**(q)|CQ;dq)	I[**v**(q1),**v**(q2)]∗(**f**(q) |CQ;**dv**)
with respect to **v**	I[**v1**,**v2**](**f**(**v**)|CV;dv)	I[**v1**,**v2**]∗(**f**(**v**)|CV;**dv**)
positive quadrant		I[M]•(**f**(**v**);**dv**)
Sectional		I[M]∗(**f**;**dV1**)

Symbols for local Ingradients

Einstein: "How does this calculus translate into Theoretical Physics?"

Breton: "We've covered many topics. We can expect that each mathematical object will need to be restricted, and once restricted will blossom into a panoply of related ideas."

Newton: "Let's start."

Breton: "We would be starting on a long journey. But the hour grows late, and tomorrow is another day."

Newton: "Agreed. We have been so involved we even skipped our lunch."

In acknowledgment the three friends arose from their Windsors, extinguished the few remaining embers in the fireplace, and left their clubhouse pensively, but with a certain anticipation for the next day.

www.ingramcontent.com/pod-product-compliance
Lightning Source LLC
Chambersburg PA
CBHW051948290426
44110CB00015B/2151
* 9 7 8 0 9 8 4 4 2 9 9 9 8 *